SCIENTIFIC INFERENCE

SCIENTIFIC INFERENCE

BY

SIR HAROLD JEFFREYS
D.Sc., F.R.S.

*Emeritus Plumian Professor of Astronomy and Experimental Philosophy
in the University of Cambridge*

THIRD EDITION

CAMBRIDGE
AT THE UNIVERSITY PRESS
1973

Published by the Syndics of the Cambridge University Press
Bentley House, 200 Euston Road, London NW1 2DB
American Branch: 32 East 57th Street, New York, N.Y.10022

New material in this edition © Cambridge University Press 1973

Library of Congress Catalogue Card Number: 71-179159

ISBN: 0 521 08446 6

First published 1931
Reissued with Addenda 1937
Second edition 1957
Third edition 1973

Printed in Great Britain
at the University Printing House, Cambridge
(Brooke Crutchley, University Printer)

CONTENTS

PREFACE TO THE THIRD EDITION

In this revision many additional relevant experimental facts are discussed. It often happens that several results are suggestive but that the clearest evidence comes later. Some relating to the special theory of relativity have extended the domain of verification greatly; and the gravitational shift of spectral lines is now shown far more clearly by the Mossbauer effect than in the companion of Sirius.

The chapter on relativity is extended. The 'clock paradox' is discussed and, I think, shown to be due to insufficient attention to methods of comparing standards.

The chapter on Statistical Mechanics and Quantum Theory is greatly expanded. A clearer explanation of entropy is given, I think, than in most works on thermodynamics. Two fundamental theorems, due to Fréchet and Cramér and Yaglom, are given. The latter appears to give a clearer reason for the importance of hermitian matrices than is usual.

My general opinion is that current accounts of quantum theory are unsatisfactory through inadequate attention to probability theory, though many of the predictions are verified. I give a tentative discussion that I hope may help somebody to make the theory more coherent.

I am grateful to Professor A. Landé for much discussion and to Dr M. Bloxham for comments throughout. My wife has given me much help with the proofs and the index. I should like again to thank the staff of the Cambridge University Press for their courtesy and patience.

HAROLD JEFFREYS

ST JOHN'S COLLEGE
CAMBRIDGE
March 1973

PREFACE TO THE SECOND EDITION

THERE have been so many developments in the theory of scientific inference since the publication of this book that I have found it necessary to rewrite the greater part of it. The general standpoint, that scientific method can be understood if and only if a theory of epistemological probability is provided, remains unaltered. Consequently I maintain that much that passes for theory of scientific method is either obscure, useless or actually misleading.

The theory is developed more fully than in the first edition. Enough is given to bring out the essential principles. Some points are treated at greater length in my *Theory of Probability*; others provide a field for further investigation. The applications are extended; in particular a new chapter on statistical mechanics and quantum theory has been added. An elementary treatment of this subject would be out of the question, but in view of the allusions to probability in its interpretations some account of the part that I think probability actually plays in it seemed desirable.

I am indebted to Professor M. H. A. Newman, Dr N. R. Hanson and Mr E. W. Bastin for advice on special points, and to my wife for comments throughout and for help with the index.

HAROLD JEFFREYS

ST JOHN'S COLLEGE
CAMBRIDGE
May 1955

PREFACE TO THE FIRST EDITION

T H E present work had its beginnings in a series of papers published jointly some years ago by Dr Dorothy Wrinch and myself. Both before and since that time several books purporting to give analyses of the principles of scientific inquiry have appeared, but it seems to me that none of them gives adequate attention to the chief guiding principle of both scientific and everyday knowledge: that it is possible to learn from experience and to make inferences from it beyond the data directly known by sensation. Discussions from the philosophical and logical point of view have tended to the conclusion that this principle cannot be justified by logic alone, which is true, and have left it at that. In discussions by physicists, on the other hand, it hardly seems to be noticed that such a principle exists. In the present work the principle is frankly adopted as a primitive postulate and its consequences are developed. It is found to lead to an explanation and a justification of the high probabilities attached in practice to simple quantitative laws, and thereby to a recasting of the processes involved in description. As illustrations of the actual relations of scientific laws to experience it is shown how the sciences of mensuration and dynamics may be developed. I have been stimulated to an interest in the subject myself on account of the fact that in my work in the subjects of cosmogony and geophysics it has habitually been necessary to apply physical laws far beyond their original range of verification in both time and distance, and the problems involved in such extrapolation have therefore always been prominent.

My thanks are due to the staff of the Cambridge University Press for their care and courtesy; also to Dr Wrinch and Mr M. H. A. Newman, who have read the whole in proof and suggested many improvements.

HAROLD JEFFREYS

ST JOHN'S COLLEGE
CAMBRIDGE
January 1931

CHAPTER I

LOGIC AND SCIENTIFIC INFERENCE

The Master said, Yu, shall I tell you what knowledge is? When you know a thing, to know that you know it, and when you do not know a thing, to recognize that you do not know it. That is knowledge. *Analects of Confucius* (WALEY'S translation.)

The fundamental problem of this work is the question of the nature of inference from empirical data so as to predict experiences that may occur in the future. An astronomer accepts without question the positions of the planets as given, for some years in advance, in the *Nautical Almanac*; a botanist is equally confident that the plant that grows from a mustard seed will have yellow flowers with four long and two short stamens. In both cases the predictions are made by way of 'scientific laws', which are based on previous instances. This type of inference is not confined to what is usually called 'science', but pervades ordinary life and even art. When I taste the contents of a jar labelled 'raspberry jam' I expect a definite sensation, inferred from previous instances. When a musical composer scores a bar he expects a definite set of sounds to follow when an orchestra plays it.

Such inferences are not deductive, nor indeed are they made with certainty at all, though they are still widely supposed to be. In relation to the botanist's inference, for instance, a logician might ask for a precise argument leading to it, and a discussion on the following lines might follow.

B. A botanical species (A) is characterized by certain specific qualities.† These specific qualities are inherited. The specific qualities of mustard include the properties that the flowers are yellow and have four long and two short stamens. Therefore the seed from a mustard plant will produce a plant with yellow flowers with four long and two short stamens.

L. I query the 'will' in your last sentence. Are you speaking of a definition of mustard, or of the behaviour of a plant that you already know to be mustard? You have many instances within your

† Lettered references are to comments at the end of the chapter.

present experience of plants that satisfied your definition of mustard and whose descendants have also satisfied it. But these are all past observations. If instead of 'will produce' you substitute 'have produced' you are simply asserting a proposition that is directly known to you. But when you say 'will produce' you are going beyond your actual knowledge.

B. But if a rule has always been true in the past it is surely reasonable to suppose that it will continue to hold in the future.

L. Some principle like that is certainly being used, but what I should like is to get it clearly stated. How do you decide what properties are specific? A particular mustard plant is 30 cm. in height and has 20 leaves. Are these qualities specific, and if so, are they inherited?

B. These qualities are not constant for a given species. To some extent they are useful in identification because some species never reach a height of more than 10 cm, and others never have more than two leaves, but they do vary even among descendants of the same individual.

L. Then for a quality to be specific, is it necessary that it is always inherited?

B. Yes, I said so.

L. How do you know that there are such qualities?

B. By experience.

L. Not entirely, I think. I grant that there are qualities that have always been found to be inherited in the past. But surely if you examined more specimens you might find instances where they fail to be inherited?

B. As a matter of fact that has sometimes happened already. We say then that either the strain was not pure and that the failures were due to Mendelian segregation, or that a mutation has occurred. Inheritance is true in a pure strain if there is no mutation. Perhaps I should have said that before.

L. Is your statement a definition of a pure strain? (B)

B. I suppose it is a working definition.

L. Then how do you know that a strain is pure?

B. The only way is to try it. I should raise a large number of seedlings from the plant, breed from them, and see whether any lines have failed to breed true with respect to the characters that I am considering.

L. But that only introduces the same problem in a different setting. Even if a strain was impure you might stop raising seedlings before you had found any that departed from type.

B. It is very unlikely. When a strain is impure we generally get a new type in about one-quarter of the cases, so if we had a few hundred specimens we should be pretty certain to detect it.

L. I am not quite sure what 'unlikely' means, though it may be possible to find a meaning for it. However, I understand that even if you had a hundred specimens, each with a one-quarter chance of having a quality, the quality might happen to miss every one.

B. You are making difficulties. We have plenty already. A friend of mine engaged in genetic experiments told me the other day that he had 15,000 plants to harvest, each of which had to be separately examined and described, and then catalogued with its pedigree. Until that is done he cannot get on to the analysis of his results at all. You appear to be asking that he should make an infinite crop before he can say anything. That is preposterous.

L. It would be preposterous to ask it, and I don't. But, if I may speak colloquially for a moment, does your answer not imply that your definition of a pure strain is a bit fuzzy at the edges? Would it be fair to say that you have an idea in your mind of what a pure strain is, but that you can never in practice know definitely whether a particular strain is pure or not?

B. You seem to be talking Plato, but I thought modern philosophers considered him out of date. As a matter of fact the geneticists have a perfectly clear definition of a pure strain. There are little bodies called chromosomes in the nucleus of each cell of a plant, and all the hereditary qualities in a plant are determined by those of the chromosomes. When two parent cells unite and subdivide, the daughter cells share the qualities from the corresponding chromosomes, and if the parent cells are alike in a quality the daughter cells must carry the same quality. A plant belongs to a pure strain if the chromosomes of all cells carry the same qualities. In botany we have rather an advantage over the zoologists, because many more plants than animals are self-fertile.

L. That is interesting. I am not sure whether I fully understand it. Do I gather that you can actually inspect the chromosomes of every cell of a plant and verify that they are alike in every cell?

B. That would be impossible—there are too many of them. But we can examine a few hundred from a plant quite easily. An ordinary microscope will show the chromosomes themselves; in recent work the electron microscope has been used, and shows up actual details in the chromosomes, which appear to carry the hereditary factors.

L. I looked at a British Flora the other day and noticed that it listed and described several thousand species of flowering plants. Has this examination by the electron microscope been carried out for every one?

B. Far from it, but it has been done for so many that we can trust the rule in general. There are some difficult cases where we were quite uncertain what were pure strains until chromosome analysis enabled us to say which were and which were not.

L. The criterion seems fairly clear, but there are still some big gaps in the application. You started by defining a quality in a plant as specific if it is inherited. But suppose I want to know whether a given quality is specific in this sense. It is not too difficult to test it up to a point, though the test does not amount to proof in the sense of deductive logic. But I began by speaking of inferences to events in the future; we are now speaking of inferences from a set of cells in a plant to all the cells in the plant, then to all in plants of the same species. I do not see that chromosomes help much in deciding what qualities are specific; to settle that you would have to use the electron microscope on every plant, and that is even harder than breeding from it. You seem to be arguing further from the data than ever; but whereas the argument before was to unexamined times, it is now also to unexamined places.

B. Still, I think that we are approaching a stage when all this can be avoided. The details that we now see in an electron microscope are not much larger than some organic molecules. Many of us have long believed that the properties of living matter can ultimately be explained in terms of physics and chemistry, and when that happens we should be able to predict all biological behaviour mathematically. In fact the physiologists have gone a long way in that direction already.

L. I don't know much about modern physics, though I have tried to understand a good many so-called popular expositions. But when you get down to molecular dimensions you are not far above atomic and electronic dimensions. Physicists now say that determinism breaks down fundamentally at these dimensions, so it seems that your hope that determinism in biology can be justified by determinism in physics must be disappointed.

B. Well, suppose I admit that we are drawing conclusions that we cannot prove deductively. Am I to give up botany for that reason? After all, we have managed to get on very well without this sort of criticism. There are, as you say, some thousands of species of plants in the British Isles, but there are some billions of individual plants. It is something that it should be possible to classify them into a few thousand species. The process of identification of a plant sometimes uses about 20 characters; and the Flora lists in each case several further characters, which are always found when the identification has been carried out. (It usually gives also some that are not always found; for instance, the number of petals may vary, but the range of variation would be given. Celandine usually has about 10 petals, but individual flowers may have from 3 to 20.)

I suppose (I was never very good at mathematics) that if you are prepared to ask 20 questions there are about a million sets of 'yes or no' answers; but we get down to sufficiently homogeneous groups long before that, and I don't think that the difference can be a matter of chance.

L. You wouldn't give it up even if I asked you to, and I don't want you to. All logicians accept scientific results. Russell spoke of applying scientific method to philosophy; Carnap goes so far as to say that logic is the syntax of the language used in science. On the other hand not many of them have first-hand experience of scientific research, and I am not sure that they get to grips with the problems of method that are important to science. They mostly get their information from semi-popular works, some of which claim to be contributions to philosophy and have been severely handled as such, and some are also strongly criticized by other scientists.

It appears that scientific method would come to a dead stop if scientists did not argue beyond their data in a way that old-fashioned logic would utterly condemn, but none of the works I know have much to say about this. I had thought that there might be some way of bringing scientific method within the scope of deductive logic, but this feature seems to be a fundamental difference.

I noticed several times in this discussion that you used such expressions as 'unlikely', 'chance', 'pretty certain', always in connexion with some problem where you appeared to be arguing beyond your data. Could you clarify them?

B. No, but surely if logic is the analysis of the language of science, it is a logician's job to do the clarification! Do you really think it important?

L. For me, it is mainly a matter of intellectual comfort. We seem to need a new sort of logic, in which proof means something different from deductive proof. I should like to do it myself, but I don't know enough about science to have much hope of covering the ground. Modern logic gets some very interesting results, especially about the foundations of mathematics, but I must admit that most ordinary mathematicians do not take much notice of them. Still, methods of modern logic may have helped to raise the standard of what most mathematicians are willing to accept as a proof. I said this to a mathematician the other day, but he replied that they had found out for themselves many cases where methods that had been widely used gave wrong answers, and that they had found out why and stiffened up the standard of proof before modern logic started. Still, I think that modern logic has done something in suggesting where they should look for mistakes. It seems to me that science needs analysis more than mathematics. For one thing, all scientists must accept mathematics; even the least mathematical botanist must be able to count stamens. Even if you often argue beyond the data, you also argue within them, so you need the elements of mathematics in addition to whatever principles you need in arguing beyond them.

B. I think the mathematics most botanists use would be accepted by anybody. But many, especially in genetics, use quite a lot of statistics. They don't understand the principles or the mathematics, but there are some books, especially by Fisher, which tell them exactly what to do, and they do it. On the whole they get on very well. The queer thing is that there are many schools of statistics, which refuse to accept one another's principles and occasionally get seriously different results in specific problems.

L. Do they tackle the problem of inference beyond the data?

B. It is hard to say. Most of them have the gift of obscurity. Many of them are attached to agricultural research stations, which give advice to farmers about the treatment of crops, so it looks as if they actually do make inferences beyond the data in expecting the farmer's treatment to produce results similar to those found on experimental plots. But I remember seeing a statement by one of them that anybody that tries to draw conclusions about future events from analysis of past ones does so at his peril. I could not see what he was driving at. The rest of the paper was too mathematical for me.

L. It looks as if this might be interesting, though the mathematics may be too hard for me too. Perhaps I might manage some of the early works. Do you know who started it?

B. They usually make some scathing remarks about a man called Bayes; I don't know who he was. And there was a Frenchman called Laplace, who seems to have just copied Bayes's mistakes.

L. What! But it can't be the same man. There was a Laplace who made some of the biggest advances in dynamical astronomy since Newton; he explained a lot of things in the motions of the planets for the first time. He must have been pretty good.

B. Better look him up and make sure. But anyhow, what can you hope to do? You say that the problem doesn't fall within logic, so how can anything be done by trying to treat it logically?

L. It is a long story. Logic is very elastic. Counting of classes goes back to very early times. The Greeks tried to apply a strict logical treatment to geometry and got an adequate theory of rational fractions, which were effectively defined in terms of comparisons of whole numbers. But they proved that $\sqrt{2}$ was not a rational fraction and suspected the same of π, though these are ratios that occur in geometry. Eudoxus made a satisfactory theory of irrational numbers, which appears in Euclid's fifth book, but mathematicians generally seemed to miss the point until Dedekind, in the early nineteenth century, showed how an irrational number can be defined (or at any rate identified, assuming it to exist) in terms of classes of rationals. Even the mysterious square root of -1 was dealt with. Some quadratic equations have no roots within the system of real numbers, but by introducing a symbol i whose square was replaced by -1 wherever it occurred in the formulae mathematicians showed that all quadratic equations could be solved in the extended system. At this stage they attached no meaning to i, but Gauss pointed out that complex numbers of the form $a+bi$ were specified by the two real numbers a, b, and consequently that all the algebra of complex numbers could be restated in terms of that of real numbers.

One feature was that to maintain the rule $i^2 = -1$ it was necessary to adopt a rather peculiar rule for multiplication of number pairs, but, given the rule, all the operations that can be carried out with real numbers can equally well be carried out with complex ones.

B. If you could do it all in real numbers, what is the point of introducing these complicated ideas at all?

L. Chiefly to save time. One equation written in complex notation expresses two in real numbers, and for those that we want it is often much simpler in form. Many results about real numbers are most easily proved by using the theory of complex ones. (C)

But the point I wanted to make was that, starting with a system that dealt entirely with positive whole numbers, mathematicians extended it step by step till they could prove properties of complex numbers, where neither member of the number pair need be rational. Of course they have gone a lot further by now.

B. You seem to be right about the elasticity of mathematics. But you spoke about logic; surely that is fixed for all time?

L. In a way, yes. But modern logic started with Peano and Frege, and the most important single work in it is still the *Principia Mathematica* of Whitehead and Russell. They tried to show that pure mathematics is a branch of logic. With a few elementary propositions of logic and the notion of membership of a class, they showed how it was possible to attach meanings to propositions about whole numbers, rationals, and real numbers and to prove enough propositions to enable the rest of the theory of real and complex numbers to be built up as had already been done by previous writers.

However, mathematicians were not altogether satisfied. Some thought that some of Whitehead and Russell's postulates were far from obvious. (D) Others had doubts about their definition of the number of a class. By definition they said that two classes were similar if their members could be paired off until both classes were exhausted, and then defined the number of a class as the class of all classes similar to that class. (A class was regarded as similar to itself.) Some mathematicians, and also some physicists, thought that they had often to work with numbers that did not satisfy the *Principia* definition, and consequently thought that the *Principia* justification of mathematics was of no use to them. Hilbert, in particular, said that what was needed was a system that would contain a set of primitive mathematical propositions, together with a set of rules that would make it possible to decide, by straightforward applications of the rules, whether any given mathematical proposition is a consequence of the primitive propositions. He also hoped that it could be proved that no contradiction exists in mathematics. However, K. Gödel showed that this programme is impossible. He showed that in any consistent system that includes arithmetic, it is possible to state a proposition that can be neither proved nor disproved within the system. On the other hand if the proposition that no pair of contradictory propositions can be proved in the system could itself be proved in the system, then a proposition that can be seen to mean the contrary could be proved in the system.

B. That sounds tough. Do you mean that in any case it is possible to deduce contradictory propositions from the postulates of pure mathematics?

L. No, in a sense it means just the opposite. There is a proposition due to Aristotle, who was nearly as old as Plato but did not often agree with him, that if in a language both a proposition and its contrary can be proved, then *every* proposition that can be stated in the language can be proved. (E) So if we can find one proposition that cannot be proved in the language, there is no pair of contradictory propositions that can be proved in the language.

B. But does not that amount to a proof of consistency, which you seemed to be saying would imply one of inconsistency?

L. The point is that to any language there corresponds what is called a metalanguage, in which we can talk *about* the language but which is not part of it. The argument I have just given is about the language and therefore in the metalanguage; we are not working within the system but looking at it from outside. As a matter of fact we usually do not say that we can prove consistency; we say that we can *see* it. The distinction does not often matter. 'Your eyes are blue' and 'it is true that your eyes are blue' appear to mean the same thing; but if 'your eyes are blue' is a statement in one language, 'it is true that your eyes are blue' is a statement about the first statement, and in logic we should have to regard it as a statement in another language.

B. It seems a delicate distinction. I remember a puzzle about a Cretan who said 'No Cretan ever speaks the truth'. Is it the same point?

L. Yes. Put it the other way round: 'Every statement made by a Cretan is false.' Whatever language this is in, it is *about* other statements, which therefore cannot possibly be in the same language. It is conceivable that a Cretan might always lie in language *A* but speak the truth in another *B*, so that his statement might be a true statement in *B* about statements in *A*. (F)

B. I don't see how you can prove that a proposition cannot be proved. There are plenty of propositions that I can't prove but other people can. Are you not going beyond your data yourself?

L. In some cases we can prove it. The system has to contain a set of rules for testing formulae. These rules are simply manipulative. If a formula can be broken up into other formulae, each of which is a postulate of the system, the formula is proved. But we sometimes find that whatever we do we always get to something that contains the formula that we are trying to test, and we are no further forward.

B. Surely logic is not just a matter of manipulating symbols? When you have written a formula down and turned over the page, how do you know that it will still be there when you have to refer to it again? Sorry to mention it, but I seem to be catching this kind of criticism from you.

L. That is a point that many logicians think it rather impolite to mention. Russell himself has said something rather similar, and so has Popper. Popper would emphatically deny that logic is nothing but the manipulation of symbols. He maintains that the common-sense idea of a proof 'this follows from that' cannot be dispensed with, and I think he would regard the symbolism only as a useful aid to memory.

B. But surely it is shaky anyhow. If memory needs symbolism to help it (and mine certainly needs a notebook) we come to using the properties of natural objects to justify logic. But the permanence of such objects is based on an argument beyond the data, which you have apparently agreed to accept, but not as part of deductive logic.

L. I must admit that I am not very clear on the point myself. Russell said that $2+2=4$ in all possible worlds, merely as a consequence of the definitions, whether there is a mind to appreciate it or not. On that basis you could hardly apply the symbolic method, since there might be no person to write the argument down.

B. You seem, all the same, to have got a long way from the Whitehead-Russell programme. All you are claiming now is that in some sense pure mathematics is consistent, which is a very different thing from saying that its results can be proved. Physicists use a lot of mathematics, but your argument suggests that they might be getting wrong results all the time without having any means of finding them out. Some biologists would rather like to believe that.

L. That is where experiment comes in. A physicist would have to make a language capable of describing his experimental data, and within that language to state postulates and rules of procedure. If these can be formulated so as to form part of a mathematical system known to be consistent, there is no further danger of inconsistency, and that is an appreciable gain. Of course there might be disagreement with experiment at a later stage, and in that case a new formulation would be necessary. Mathematics can display connexions between scientific statements; it does not prove the statements by itself.

B. That makes sense. I was always a bit tired when people said that something in science had been proved by mathematics. But I am envious of the exact sciences. In biology we always have to deal with samples that vary over a range; we never get exact agreement however hard we try. I don't think mathematics would help us much even if we knew more about it than we do.

L. I am not sure that the difference is as great as all that. Physicists are always claiming to have made more accurate measurements than anybody has made before; which implies that previous measures were not absolutely accurate, and the new ones may not be. They say, for instance, that they have determined all atomic weights more accurately than the best chemical methods ever did. So I think 'exact science' is a rather unfortunate expression; the physical and biological sciences may differ in degree with respect to the accuracy with which one can predict an observation, but I see no reason to suppose that there is any difference in principle.

B. But whether science can be exact or not, mathematics is. If all observations are inaccurate, how can anybody say anything at all about them by mathematics?

L. Take the *Nautical Almanac*. If the values of the positions and velocities of all the planets are known to have lain within specified ranges at 1900·0, mathematics can calculate values for other times corresponding to any set of values at 1900·0, and hence a set of ranges within which the positions at each later time must lie. But even within mathematics itself results are often given with a range of uncertainty. For many purposes it may be enough to know the answer within a certain range, and there is no point in carrying the calculation to more figures than we really need. A 4-figure table of sines may contain 5401 entries, all of which are right within 0·00005, but only three (for 0°, 30°, and 90°) are exact. However, you mentioned that a plant of a given species does not always have the same number of petals, and you can count the petals exactly. So there must be more in it than that.

B. Are you hoping to construct a determinist scheme after all, which would lay down the exact number of petals that every individual flower must have?

L. It would be nice if we could, but even if we could we have not done so yet, and we need to know what to do in the meantime. There are many societies and journals that discuss the philosophy of science, but when it comes to causality they never appear to get anywhere. Scientists take part in the discussions, and presumably know more about it than the philosophers, but they differ among themselves even more violently than the philosophers do.

B. I dare say some that don't come to the discussions express themselves more violently still.

Statisticians get their first ideas from the throwing of dice; this is a case where the number of pips is bound to be an exact number, but is not predictable. That is something like the state of affairs in the number of petals of a flower. I suppose that if we had a machine that could reproduce the conditions of throwing sufficiently accurately we could always predict the result.

L. Yes, but that is not all. With dice you do at least know the number of faces and what numbers are on them before you begin. In your celandine example you know only that certain numbers of petals have occurred in actual flowers; you would be saying more than you know if you said that there must always, say, be more than 3 and less than 20. A theory that is to be of any use must include the behaviour of dice, but must also enable us to use the observations to say something about systems where the number of possibilities is unknown and even infinite.

B. That sounds a colossal undertaking! But do you really think it possible?

L. I don't know. But scientists are making this sort of inference all the time, without knowing much about how or why they do it. So I am sure that somebody should try. It would certainly need a lot of mathematics, and a good deal of knowledge of actual scientific procedure. The difficulty about expressing the ideas mathematically does not appear insuperable, considering how successful mathematics has been in extending its scope at various times.

B. It does look as if nobody knows all the logic, science, and mathematics needed to finish the job. But what good would come of it?

L. I suppose some clarification of ideas. I notice that scientists sometimes get into acrimonious discussions on how to handle their data. It might be possible to find rules that would help them to come to a decision with less trouble.

B. I am not sure that some of them do not like acrimonious discussions for their own sake.

The main point of the foregoing discussion is that inference from past observations to future ones is not deductive. The observations not yet made may concern events either in the future or simply at places not yet inspected. It is technically called induction. This statement is at least as old as Plato, and was emphasized by Hume at considerable length. There is an element of uncertainty in all inferences of the kind considered. Most scientists have ignored Hume's difficulty. Some philosophers have overemphasized it by declaring all sensory information unreliable. Others have claimed to answer it, but all the answers depend on elementary mistakes; I shall discuss some of them later. I should mention at once, however, one consequence of philosophers' use of popular and semi-popular books as sources of scientific information. These books necessarily make scientific problems look easier than they are. Where science produces a quantitative law, such as Pythagoras's theorem, the observations do not satisfy the law exactly. (Plato knew this.) In the problems of plant breeding discussed above, the botanist begins as he might in an elementary lecture, but finding his questioner critical he gradually volunteers various bits of information from more advanced parts of the subject, and it emerges not only that laws are not deductively proved, but that many laws previously accepted have already been found to have exceptions. A common argument for induction is that induction has always worked in the past and therefore may be expected to hold in the future. It has been objected that this is itself an inductive argument and cannot be used in support of induction. What is hardly ever mentioned is that induction has often failed in the past and that progress in science is very largely the consequence of direct attention to instances where the inductive method has led to incorrect predictions. *Scientific progress never achieves finality*; *it is a method of successive approximation.* The botanist's experience is an instance of this, but he had not recognized the fact clearly, and appears to become entangled in contradictions in consequence. What he actually does is to state his laws in the most general and simple form possible, and modify them as later experience indicates. I propose to show not only that this process is free from self-contradiction, but that it is the only method possible.

NOTES

A. It should be said that B's first description of a species would be modified by many biologists, who would say that no set of characters is always present in a given species. All characters are subject to some variation. Whether two plants belong to the same species is to be judged by the presence of a number of characters in common out of a certain set. Such a criterion introduces the practical difficulty that, to know whether two individual plants belong to the same species, we could not be satisfied with this vague statement; we must specify the characters and say just how many of them are required to be present in both plants. Then L's comments would need little change.

Another criterion sometimes proposed is based on fertility. Species are grouped into genera, and subdivided into varieties. The suggestion is that (1) different varieties of the same species produce fertile hybrids, (2) different species within the same genus produce infertile hybrids, (3) plants of different genera cannot produce hybrids. The good point of this proposal is that it provides an experimental criterion; the current divisions between species are somewhat arbitrary. Against it are, first, the difficulty of application (impossibility for extinct species and those that always reproduce vegetatively) and secondly, it may happen that two plants that do not cross can both be fertilized by a third. This is true, for instance, with some cultivated fruits.

These two procedures differ from B's, one starting more vaguely, the other more precisely, but neither of them escapes the difficulties, and would be criticized by L at least as severely as B's formulation.

B. The discussion of the attempt to define a pure strain affords an instance of the common assertion that scientific laws are disguised definitions. An extreme form is Russell's remark,† 'Things are those series of aspects that obey the laws of physics'. This would amount to saying that the laws of physics are chosen first, and we then look for something to satisfy them. This inverts the actual procedure, which has been painfully learnt. Mechanics struggled for centuries to reconcile motion with the rule that velocity is proportional to force, chemistry for about one century to reconcile its phenomena with the phlogiston theory. The attitude propounded by Russell was the chief cause of delay in constructing the Galilean theory and the laws of chemical combination. It should be said that neither pre-Galilean mechanics nor the phlogiston theory was absurd in itself.‡

A leading part in the modern development of theories of scientific method was played by Ernst Mach, who insisted on the direct discussion of observations and regarded scientific laws as the description of observations in the most economical way. Like Russell, he is best known for his less fortunate remarks, but in his case they are more often believed. The expression 'economy of thought' led to the comment that the best way of economizing thought is not to think. The fact is, of course, that to describe observations in the most compact way needs a great deal of thought. Mach missed the point that to describe an observation that has not been made yet is not the same thing as to describe one that has been made; consequently he missed the whole problem of induction.

† *Our Knowledge of the External World* (1914), p. 110. Russell's less fortunate remarks have a way of being the most often quoted, even after he has abandoned them. But both they and the objections to them are always interesting.

‡ H. Butterfield, *The Origins of Modern Science* (1950), chapters I and II.

This was taken up by Karl Pearson in *The Grammar of Science*,†
which is still the outstanding work on the subject.

C. Take the identity, for complex numbers $z = x + iy$:

$$(z_1 + z_2)(z_1 + z_2) = z_1^2 + 2z_1 z_2 + z_2^2.$$

If we express this in real numbers we get the pair of identities

$$(x_1 + x_2)^2 - (y_1 + y_2)^2 = (x_1^2 - y_1^2) + 2(x_1 x_2 - y_1 y_2) + (x_2^2 - y_2^2)$$

$$2(x_1 + x_2)(y_1 + y_2) = 2x_1 y_1 + 2(x_1 y_2 + x_2 y_1) + 2x_2 y_2.$$

Either of these separately is more complicated than the original
identity for complex numbers.

D. The disputed axioms in *Principia Mathematica* are the axiom
of reducibility, the axiom of infinity, the theory of types, and the
multiplicative axiom. There is a discussion by F. P. Ramsey.‡
The theory of types was created to deal with a difficulty found by
Russell. The class of all classes is apparently a class, but the class
of all dogs is not a dog. So there are some classes that are not
members of themselves. Is the class of all classes that are not
members of themselves a member of itself or not? Suppose it is;
it follows that it is not. Suppose it is not; then it is. Russell
dealt with this by saying that it is meaningless to say that a class is
a member of itself; such a statement is neither true nor false.
Similarly a proposition cannot be about itself.

Individuals are said to be of (logical) type I. Properties of
individuals are of type II. (Classes can be treated similarly to their
defining properties.) Properties of properties of individuals are of
type III, and so on. A property of type n can meaningfully apply
only to properties of types $n - 1$ and lower. This classification avoids
Russell's contradiction. It is possible that Russell's criterion is
unnecessarily strict, and later writers have proposed various modifications of the theory of types, but some such theory is necessary in
any case.

The axiom of reducibility, the multiplicative axiom, and the
axiom of infinity have the common feature that they look as if they

---

† First published in 1892. The first and second editions were in two parts,
dealing with the physical and biological sciences. The third edition (1911)
brought the physical part up to date, but the biological part was never revised.
The edition in Everyman's Library contains selections from both parts.

‡ *The Foundations of Mathematics* (1931); *Proc. Lond. Math. Soc.* (2) **25**,
1925, 338–84. See also Russell, *Introduction to Mathematical Philosophy* (1919);
W. V. Quine, *Mathematical Logic* (1940).

depended in some way on the constitution of the actual world, and if so, they should not form part of logic or pure mathematics, which aim at being true in all possible worlds. I shall not discuss the axiom of reducibility, which is not used in Ramsey's modified theory. The axiom of infinity says that if a class of number n exists, then a class of number $n + 1$ exists. If it was false we could still have a restricted theory of whole numbers and rational fractions, but could not have the theory of real numbers, and the whole of higher analysis would disappear. It might be argued that in that case we should not need higher analysis. But some results about whole numbers have been proved only by means of higher analysis; and the sacrifice would be considerable.

The multiplicative axiom (treated with great reserve by Whitehead and Russell) says that for any set of classes there is a property that is possessed by precisely one member from each. In some cases we can point out such a property, and do not need the axiom. The axiom is used in cases where we cannot point out a suitable property, but one may nevertheless exist. It can be proved from the other axioms when the set is finite, but not when it is infinite. Gödel has proved that it is consistent with the other axioms. Its denial would lead to curious consequences. A man's boots can be distinguished as right and left; his socks cannot. Suppose he has $\aleph_0$ pairs of boots and $\aleph_0$ pairs of socks. This means that there are rules that match each pair of boots and each pair of socks with a positive whole number, and conversely. Has he as many boots as socks? Apparently we cannot say. For to compare them we should need a rule that would say, for every pair of socks, which is to be associated with the right boot and which with the left; and if the multiplicative axiom is false there may be no such rule.

This difficulty can be avoided in this particular case. Take the centre of mass of each sock, and for every pair of socks take the plane consisting of points equidistant from the centres of mass. These planes do not fill all space. Take a point not on any of them. With each right boot associate the nearer sock, with each left boot take the further sock. But there is no reason why such a trick should always work. In practice mathematicians use the multiplicative axiom wholesale while hardly noticing it.

Its acceptance, on the other hand, leads to a curious consequence. The Lebesgue definition of the measure of a set of points on a line from o to 1 is designed so that, however complicated the definition of a set may be, the measure of the set and that of the set consisting of the rest of the points from o to 1 shall have sum 1. It achieves this object for many sets where the older theories did not. The more general additive property of Lebesgue measure is one of the principal advantages claimed for the Lebesgue definition. But if the multiplicative axiom is assumed we can prove the existence of sets that do not possess the additive property—the definition is less successful than had been hoped.

E. A little symbolism helps. If p, q, r are propositions, we write '$\sim p$' for '(not p)', '$p.q$' for '(p and q)', '$p \vee q$' for '(p or q)'. '$p \vee q$' includes the case where p and q are both true. We write also 'p ent q' for '(p entails q)', i.e. q follows logically from p.

If q ent r, then $p.q$ ent $p.r$. Then

$$p \text{ ent } (p \vee q);$$

$$(\sim p.p) \text{ ent } (\sim p.(p \vee q))$$

$$(\sim p.(p \vee q)) \text{ ent } q.$$

Hence $(\sim p.p) \text{ ent } q,$

where q may be any proposition in the language whatever.

McTaggart is said to have denied the conclusion, saying, 'If twice 2 is 5, how can you prove that I am the Pope?' G. H. Hardy answered 'If twice 2 is 5, $4 = 5$. Subtract 3; then $1 = 2$. But McTaggart and the Pope are two; therefore McTaggart and the Pope are one.'

Gödel's theorem is based on the use of three languages, which I shall denote by L (logical), N (numerical) and M (the meta-language). L contains numerical signs, which are interpreted in M as meaning the corresponding numbers in the usual way. It also contains expressions for the ordinary logical terms and mathematical functions. In any formula of L of finite length the symbols are in a definite order, and the places can be indicated by prime numbers 3, 5, 7, 11, ... A symbol of L is associated with a number q, and the statement that the symbol with number q occurs in the place associated with the prime p is indicated by the factor p^q. (Differences of type are taken into account in the choice of q.) Then in N any formula of L is expressed by an actual number, the product of p^q over all p. This is the Gödel number of the formula.

Thus every formula in L can be expressed by a Gödel number (G-number for short). By factorizing the G-number we can recover the original formula of L in a finite number of steps. If a Gödel number was infinite the factorization would of course be impossible. We are therefore concerned with proofs in a finite number of steps. In general a number in L is represented by a different number in N.

The system of L is stated in L, and correspondingly for N.

In M we can speak of both L and N.

If in an expression $F(u)$ of L, u can take more than one numerical value, we can consider first $f_0, f_1, f_2 \dots$ which are the G-numbers of $F(0)$, $F(1)$, $F(2) \dots$; and hence a mathematical function $f(n)$, which for each n takes the value f_n. We can also consider a single number g, which is the G-number of the formula $F(u)$ of L, where u is now just one of the symbols of L.

Now define a function $s(x, y)$ as follows. If m is the G-number of $F(u)$, then $s(m, n)$ is the G-number of $F(n)$. Also define $B(m, n)$ by 'm is the G-number of a proof in L of the proposition whose G-number is n'.

The sign $\exists$ is defined by saying that $\exists x . \phi x$ means 'there is an x with the property ϕ'.

Let $G(u)$ be $\sim(\exists x . B(x, s(u, u)))$ and let p be the G-number of $G(u)$. Then p in N specifies what function in L $G(u)$ is, and $s(p, p)$ is the G-number of $G(p)$. Also $G(p)$ is $\sim(\exists x . B(x, s(p, p)))$. But we see (in M) that this means '$G(p)$ is not provable (in L)', and $\sim G(p)$ means $\exists x . B(x, s(p, p))$, that is, '$G(p)$ is provable'.

Thus $G(p)$ is provable in L if and only if it is false, which looks awkward. But we can say more than this. The rules of L have counterparts in N and either can be derived from the other by the rules of translation; thus for Carnap's language LII the primitive sentences are deduced from true propositions of arithmetic. Hence if arithmetic is consistent, LII is consistent. We accept the consistency of L. Now if $\sim G(p)$, $G(p)$ is provable in L; that is, there is a proof in L that $G(p)$ is not provable in L. But if $G(p)$ is provable in L there is a proof that $G(p)$ is provable in L—namely, by exhibiting the proof of $G(p)$. Hence if $G(p)$ is false there are proofs in L that $G(p)$ is both provable and not provable in L, and this is impossible since L is supposed consistent. We see that this argument is in M.

It follows (in M) that $G(p)$ is true, that is, $G(p)$ is not provable in L. But if this was provable in L we should have 'there is a proof in L that $G(p)$ is not provable in L'; that is, there is a proof in L of $G(p)$, and again we have a contradiction. Hence if L is consistent neither $G(p)$ nor $\sim G(p)$ is provable in L; $G(p)$ is an undecidable proposition.

Thus in any consistent logical system with a finite number of signs and axioms that includes arithmetic it is possible to state a proposition that cannot be proved or disproved by means of the rules of the system.

The above argument is an elaboration of the sketch on pp. 39–41 of the translation. The actual theorem needs attention to several other points of detail and is not completed till p. 61. There have been many attempts at shorter proofs than Gödel's, but all have been found unsatisfactory. One complication is that to prove a theorem $f(r)$ for all numbers r up to n we have to prove all of $f(0), f(1), f(2) \dots f(n)$. For finite n a Gödel number can be assigned to this. But extending n to infinity would produce an infinite Gödel number. This can be avoided by introducing a symbol $(r)f(r)$ meaning '$f(r)$ for all r'. This needs a single Gödel number, but then consistency requires a postulate that we cannot have '$f(r)$ provable for every finite r and $\sim (r)f(r)$ provable'. This would not worry most people: it is called ω-consistency. It is in the M language and not the L language, and therefore has no G number. It avoids the possibility that $\sim (r)f(r)$ might be provable and yet separate proofs of $f(r)$ might exist for different finite r. Simple consistency would say that we cannot have '$\sim (r)f(r)$ and $(r)f(r)$ both provable' (by any method). Rosser gives an undecidable proposition that does not assume it; his proof is in Kleene's book.

It is necessary for a complete understanding to distinguish between a formal and an informal proof. A formal proof consists entirely of symbols manipulated according to rules stated in symbols. Interpretation comes only before and after it. The above argument uses ordinary language on the way and is therefore informal. We shall see later, however (p. 98), that the final step, that of detaching the final result and asserting it as a separate proposition, can never be formalized.

The consistency of arithmetic itself has been studied by G. Gentzen. We saw above that $G(p)$ is undecidable in L but is seen in M to be true. Gentzen produces a hierarchy of systems beginning with arithmetic, and shows that each contains an undecidable proposition decidable in a higher one; and if we are willing to accept an infinite number of systems the consistency of arithmetic can be proved.†

F. A more direct example would be 'I am lying'. Apparently if it is false it is true, and if it is true it is false. But a man could say the words in the given order, and it looks as if we have a contradiction in any case. The answer given by the theory of types would be that the statement, being about itself, is meaningless. The answer given by L would be that it cannot be about itself, and we do not know what statement in some other language it is about. (A similar answer has been given by the physicist Bridgman.)

An interesting example is the following. In a certain town there is one barber. He shaves all the men that do not shave themselves and only those men. Who shaves the barber? If he does not shave himself, he shaves himself. If he shaves himself, then since he shaves only men that do not shave themselves, he does not shave himself. We have a contradiction in either case. This looks like the kind of paradox that is solved by the theory of types; but I think that it is not. The theory of types would say that we can define exclusive classes C_1, 'men in the town that shave themselves', and C_2 'men in the town that do not shave themselves'. Then 'the barber shaves all men of C_1', if he shaves himself, would be a proposition about all men of C_1 and therefore about 'the barber shaves himself'. Similarly if he does not shave himself, 'the barber does not shave any man of C_2' would be about 'the barber shaves himself' and therefore in either case the proposition would be meaningless.

I think that there is a fundamental difference between this problem and 'I am lying'. It is perfectly possible for a man to say 'I am lying'; the difficult thing is to see what he means. But it would be impossible to satisfy the conditions proposed in the problem of the barber. Suppose we define C_1 and C_2 to be exclusive and to include every man in the town. Then the barber is in C_1 or

† K. Gödel, *Mh. Math. Phys.* **37**, 1930, 349–60; **38**, 1931, 173-98. English translation 'On formally undecidable propositions...', by B. Meltzer, with introduction by R. B. Braithwaite, Oliver and Boyd, 1962. R. Carnap, *The Logical Syntax of Language* (1937). G. Gentzen, *Math. Ann.* **112**, 1936, 493–565. S. C. Kleene, *Introduction to Metamathematics* (1952). A recent book by S. W. P. Steen, *Mathematical Logic* (1972), treats many of the problems at greater length.

he is in C_2. If he is in C_2 he does not shave himself and therefore it is false that he shaves all the men that do not shave themselves (and as there is no other barber he presumably has a beard). If he is in C_1 he shaves himself and it is false that he shaves only those men that do not shave themselves. The point is that it is perfectly possible to determine by actual examination of the inhabitants whether the conditions stated in the problem are satisfied; and such examination will show that they are not. The conditions are in fact genuinely self-contradictory, and no theory of types can resolve the contradiction. It is too easily assumed that any question has an answer; in science the framing of the correct question is often the most difficult part of the problem. In the problem of the barber the question itself assumes impossible conditions.

Dr Bloxham points out to me that the argument can be stated in terms of the *Principia* definition of material implication. In this sense 'p implies q' means simply '$\sim p \vee q$'. It differs from entailment, since it does not require q to be deducible from p; it is true if p is false, and then gives the 'peculiar proposition' that a false proposition implies any other proposition. Then the conclusion above, that the data are inconsistent, can mean simply that the data are false, which is what I say.

CHAPTER II

PROBABILITY

Oh, it ain't gonna rain no mo', no mo',
It ain't gonna rain no mo'!
How in the hell can the old folks tell?
'Tain't gonna rain no mo'!
MESSRS LAYTON AND JOHNSTONE

2·0. In the first chapter I have tried to present the problem of scientific inference by means of a dialogue between a botanist and a modern logician, both of sufficient intelligence to see that there is a problem, but both unaware, as most scientists are, that any serious efforts have been made to solve it. The methods of the modern schools of logic do, as a matter of fact, help us to describe the problem more clearly than was possible when the *Grammar of Science*, or even the first edition of this book, was published. When we make an inference beyond the observational data, we express a logical relation between the data and the inference. This relation is in a generalized logic, not in deductive logic. It does not claim that the inference is deductively proved or disproved from the data. It assesses the support for the inference, given the data, but an essential feature is that this support can be of many different degrees. A thousand instances of an event happening, with no exception, in given circumstances, are better evidence than one instance that it will happen the next time the circumstances occur. This relation between a set of data and a conclusion is called *probability*, and the subject is essentially what is now called a many-valued logic.

If we like there is no harm in saying that a probability expresses a degree of reasonable belief.† This expression helps us to understand that the main object of the theory is to tidy up the process of scientific inference, which we shall regard as reasonable on the whole, though there are instances of actual procedure that we shall find our-

---

† Braithwaite says that reasonable belief is an all-or-none reaction. This would have been regarded by W. H. R. Rivers as a definition of instinctive belief; in any case such beliefs are precisely what I call unreasonable.

selves compelled to regard as unreasonable and not covered by the theory. If we accept Bertrand Russell's statement that logic exists independently of thought, we should not speak of the theory as concerned with any kind of belief. But in practice logic would not be of much use without a mind to appreciate it. 'Degree of confirmation' has been used by Carnap, and possibly avoids some confusion. But whatever verbal expression we use to try to convey the primitive idea, this expression cannot amount to a definition. Essentially the notion can only be described by reference to instances where it is used. It is intended to express a kind of relation between data and consequence that habitually arises in science and in everyday life, and the reader should be able to recognize the relation from examples of the circumstances when it arises. Deduction cannot be defined either.

The fact that confirmation may have many degrees, ranging from deductive proof to deductive disproof as extreme cases, suggested to the early workers that it could be expressed by means of numbers over a finite interval. A practice in symbolic logic enables us to state this in another way. In symbolic logic a special mark $\vdash$ is prefixed to propositions that are asserted to be true. This mark is not a part of the proposition; it distinguishes propositions that are asserted from those that are only being considered. If a proposition is of the form not-p and is asserted, this is equivalent to denying the proposition p. A completely equivalent procedure would be to use a figure $\mathbf{1}$ (in a different type fount from any used in the proposition) as the sign of assertion, and $\mathbf{0}$ to express denial. Thus where Whitehead and Russell write the laws

$$\vdash (p \lor \sim p); \quad \vdash \sim (p \cdot \sim p)$$

(one of p and its contrary is true; p and its contrary are not both true) we could agree to express the same statements by

$$\mathbf{1} \; p \lor \sim p; \quad \mathbf{0} \; p \cdot \sim p.$$

Then intermediate degrees of confirmation might be expressed by using proper fractions in place of $\mathbf{0}$ and $\mathbf{1}$. Such a method, however, would be cumbrous in probability theory. In ordinary logic a fixed set of postulates is given at the start, and all propositions asserted later are consequences of this set, which need not be stated afresh every time. In probability theory both the data and the proposition considered are subject to alteration, and it is therefore necessary to keep the data explicit. This is achieved by writing the relation in the form

$$P(q \mid p) = a$$

(read 'the probability of q, given p, is a'), where a is the number that expresses the degree of confirmation.

We have not yet proved that degrees of confirmation can be expressed by numbers in this way. In any case they are not actually numbers; all we are saying is that any statement about them can be replaced, by assigned rules, by a statement about numbers, and can be recovered from this statement by applying the same rules backwards. Our fundamental hypothesis is that degrees of confirmation of different propositions, on the same data, can be put in order, so that our fundamental type of comparison is of the form 'on data p, q is more probable than r'. If q is more probable than r, we say that r is less probable than q; if neither is more probable than the other, we say that they are equally probable. Then our fundamental axiom† is

Axiom 1. On data p, if q and r are two propositions, either q is more probable than r, or q is less probable than r, or they equally probable; and not more than one of these statements is true.

Axiom 2. The relation 'more probable than' is transitive; that is, if, given p, q is more probable than r and r is more probable than s, then q is more probable than s.

Since for real numbers a, b, c, either $a < b$, $a > b$, or $a = b$, and if $a > b$, $b > c$, then $a > c$, these axioms are consistent with putting degrees of confirmation in one-one correspondence with a set of real numbers so that the larger number corresponds to the higher degree of confirmation. This is our first step in the assignment of the numbers. So far, however, they might be assigned in an infinite number of ways. The actual choice depends on a further axiom. We need a word to express the statement that q and r cannot both be true given p, that is, p entails $\sim(q.r)$. We say in this case that q and r are *exclusive* given p.

Axiom 3. On data p, if q, r are exclusive, and q', r' are exclusive,

† The axioms are given more fully in my *Theory of Probability* and still more fully in Carnap's *Logical Foundations of Probability*.

and if q, q' are equally confirmed by p, and r, r' are equally confirmed by p, then q ∨ r, q' ∨ r' are equally confirmed by p.

This axiom permits us to adopt the addition rule:

If on data p, q and r are exclusive, then the numbers corresponding to the degrees of confirmation of q, r and q ∨ r given p satisfy the rule

$$P(q \vee r \,|\, p) = P(q \,|\, p) + P(r \,|\, p).$$

Then axiom 3 will say that if $P(q \,|\, p) = P(q' \,|\, p), P(r \,|\, p) = P(r' \,|\, p)$, and if $\sim (q.r), \sim (q'.r')$ are entailed by p,

then
$$P(q \vee r \,|\, p) = P(q' \vee r' \,|\, p).$$

An immediate consequence is that if, given p, all pairs from $q_1, q_2, ..., q_n$ are mutually exclusive

$$P(q_1 \vee q_2 \vee ... \vee q_n \,|\, p) = P(q_1 \,|\, p) + P(q_2 \,|\, p) + ... + P(q_n \,|\, p).$$

We have stated that the extreme degrees of confirmation are proof and disproof. Then given p, all propositions contradicted by p must have the same probability, say α, and all entailed by p must have the same probability, say β. Suppose then that $q_1, q_2 ... q_n$ are all equally confirmed by p and mutually exclusive, and also that they are *exhaustive*, that is, that at least one of them must be true given p. (If the members of a set are both exclusive and exhaustive, precisely one of them must be true, given p.) Let $P(q_r \,|\, p) = \gamma$ for all r from 1 to n. Then for any m

$$P(q_1 \vee q_2 \vee ... \vee q_m \,|\, p) = m\gamma \tag{1}$$

and for $m = n$
$$P(q_1 \vee q_2 \vee ... \vee q_n \,|\, p) = n\gamma = \beta. \tag{2}$$

Hence $\gamma = \beta/n$,
$$P(q_1 \vee q_2 \vee ... \vee q_m \,|\, p) = m\beta/n. \tag{3}$$

Further, we can take $m = 0$, which would say that none of the q_r is true given p. This is impossible given p, and therefore the number attached to it is α. Hence
$$\alpha = 0.$$

More directly, take $n = 2$ and let q_1, q_2 be two propositions neither of which can be true given p. Then $q_1 \vee q_2$ cannot be true given p and therefore

$$P(q_1 \vee q_2 \,|\, p) = P(q_1 \,|\, p) = P(q_2 \,|\, p) = \alpha. \tag{4}$$

But $q_1 . q_2$ cannot be true given p; hence by the addition rule

$$P(q_1 \vee q_2 \,|\, p) = P(q_1 \,|\, p) + P(q_2 \,|\, p) = 2\alpha \tag{5}$$

and therefore $\alpha = 0$.

Consistently with the addition rule, therefore, the admissible numbers to express degrees of confirmation range from 0 to some positive number β, which we usually take to be 1. If we took β to have some other value, all the numbers would simply be multiplied by β and the addition rule would still hold.

2·1. There is nothing in the theory so far to fix bounds to n and m, and we can certainly *consider* any integral values for them. Hence the theory so far permits any rational proper fraction to be a probability. Further, if we should meet a proposition q whose degree of confirmation given p cannot be expressed by a rational proper fraction, we could consider the classes of degrees of confirmation expressed by rational fractions that are respectively less and more strongly confirmed by p than q is. These classes exist by axiom 1. They are separated by a real number (irrational); and if this is taken for $P(q \mid p)$ all the axioms are still satisfied. Hence any degree of confirmation can be associated with a real number a satisfying $0 \leqslant a \leqslant 1$.

2·2. So far we have considered only degrees of confirmation of different propositions, keeping the data fixed. We have also to consider the consequences of varying the data, and need a further axiom since we introduce new subject matter. A little inspection shows some properties that such an axiom must have if it is to be acceptable. We begin by considering the probability of $q.r$ given p, where q and r are in general not taken to be exclusive. Clearly q and r together cannot be more strongly confirmed than either separately, since q is equivalent to $(q.r) \vee (q. \sim r)$, and $q.r$ and $q. \sim r$ are exclusive. Hence

$$P(q \mid p) = P(q.r \mid p) + P(q. \sim r \mid p) \geqslant P(q.r \mid p). \qquad (6)$$

Then if $P(q \mid p) = 0$, we have $P(q.r \mid p) = 0$, and similarly if $P(r \mid p) = 0$, $P(q.r \mid p) = 0$.

Further, if q and r are both entailed by p, $q.r$ is entailed by p, and

$$P(q \mid p) = P(r \mid p) = P(q.r \mid p) = 1. \qquad (7)$$

These relations suggest that

$$P(q.r \mid p) = P(q \mid p)P(r \mid p) \qquad (8)$$

since this holds in all the cases considered. However, it is easily seen to be false by taking a case where r is $\sim q$ and $P(q\,|\,p) = \frac{1}{2}$. Then $P(r\,|\,p) = \frac{1}{2}$,

$$P(q.r\,|\,p) = P(q.\sim q\,|\,p) = 0; \quad P(q\,|\,p)P(r\,|\,p) = \tfrac{1}{4}. \qquad (9)$$

In fact it is clear that $q.r$ is impossible, given p, not only if q and r are separately impossible given p, but also if r is impossible given $p.q$. This condition can occur, as we have just seen, if r is simply $\sim q$. This suggests that the rule should be

$$P(q.r\,|\,p) = P(q\,|\,p)P(r\,|\,q.p). \qquad (10)$$

This clearly leads to no contradiction in the case just considered, for

$$P(q\,|\,p)P(\sim q\,|\,q.p) = \tfrac{1}{2}.0 = 0. \qquad (11)$$

(10) is therefore satisfactory so far, and is clearly the simplest relation that can be satisfactory. We therefore adopt it in general. It cannot be proved from the other axioms, because these have said nothing about comparing probabilities on different data, but it is worth while at this point to examine the question of consistency.

We saw in Chapter 1 that a contradiction $p.\sim p$ entails every proposition in the language. Let us use β again to denote entailment. Then for any q we have apparently

$$P(q\,|\,p.\sim p) = \beta, \qquad (12)$$

and also $\qquad\qquad P(\sim q\,|\,p.\sim p) = \beta, \qquad (13)$

and therefore, by the addition rule,

$$P(q \vee \sim q\,|\,p.\sim p) = 2\beta. \qquad (14)$$

This contradicts the rule that no probability exceeds β. Thus all our axioms need elaboration by a statement that the data must not be self-contradictory (like those of the puzzle of the barber). In science we are not interested in inferences from self-contradictory data; a contradiction among our data, if there is one, would be looked for by deductive methods and the data would be modified accordingly. (There could of course be no contradiction among observational data, but there might be one between two hypotheses or between the observational data and some hypothesis whose consequences are being examined.)

We are now in a position to go some way towards establishing consistency, assuming the consistency of pure mathematics.

Suppose that there is some general datum that is included in all experience; it might for instance be the rules of pure mathematics. Denote this by H. If proposition p has a positive (non-zero) probability on data H and q is any other proposition, assume that $P(q \mid p.H)$ satisfies

$$P(q \mid p.H) = P(p.q \mid H)/P(p \mid H). \tag{15}$$

This is true if we maintain the product rule; but we are not assuming the rule in general and this is taken to be simply a rule for calculating probabilities on other data assuming that we have them on H. We do assume that the addition rule holds on data H. Then we always have

$$0 \leqslant P(q \mid p.H) \leqslant 1,$$

and if q_1, q_2 are exclusive given $p.H$ (and therefore $q_1.p$ and $q_2.p$ given H) the definition gives

$$P(q_1 \vee q_2 \mid p.H) = \frac{P((q_1 \vee q_2).p \mid H)}{P(p \mid H)} = \frac{P(q_1.p \mid H)}{P(p \mid H)} + \frac{P(q_2.p \mid H)}{P(p \mid H)}$$

$$= P(q_1 \mid p.H) + P(q_2 \mid p.H). \tag{16}$$

Hence the addition rule is satisfied on data $p.H$.

Again, consider two sets of alternatives q_i ($i = 1$ to m), r_j ($j = 1$ to n). We have by definition

$$P(q_i.r_j \mid p.H) = P(q_i.r_j.p \mid H)/P(p \mid H), \tag{17}$$

and if $P(q_i.p \mid H) > 0$

$$P(r_j \mid p.q_i.H) = P(q_i.r_j.p \mid H)/P(p.q_i \mid H) \tag{18}$$

$$P(q_i \mid p.H) = P(p.q_i \mid H)/P(p \mid H). \tag{19}$$

Hence by multiplication

$$P(q_i.r_j \mid p.H) = P(q_i \mid p.H)\,P(r_j \mid p.q_i.H). \tag{20}$$

Hence the product rule is true if other data are combined with $p.H$. Hence if probabilities on data H are assigned so as to satisfy the addition rule, probabilities on any data can be assigned provided the data themselves have positive probabilities on H; and these probabilities will satisfy the addition and product rules.

These two results together show that, for consistency in the application of the addition and product rules,

(1) it is *necessary* that no probabilities considered shall be on self-contradictory data;

(2) it is *sufficient* that all considered data should have positive probabilities on the general datum H.

The latter condition is far from necessary. The immediate point is that 'q has probability 0 given p' is not the same thing as 'p entails $\sim q$'. We shall have examples of this later. (p entails $\sim q$) entails $P(q\,|\,p) = 0$; the converse is not true. Suppose that a measurable quantity x, on data p, must lie between 0 and 1 and that the probability that it lies in any interval is proportional (and therefore equal) to the length of the interval. What is the probability on p that x is precisely $\frac{1}{2}$? Clearly 0; but $\frac{1}{2}$ is not an impossible value. Thus $P(q\,|\,p) = 0$ does not entail 'p entails $\sim q$' when an infinite set of alternatives are under consideration. Similarly $P(q\,|\,p) = 1$ does not entail 'p entails q'. It may therefore be held that the probability scale is not sufficiently general, since the degree of confirmation of q given p may be held to be different if $P(q\,|\,p) = 0$ according as p does or does not entail q. We shall often use the expressions 'q is impossible given p' for 'p entails $\sim q$' and 'q is certain given p' for 'p entails q'. Then a usage common in pure mathematics suggests that we should use 'q is almost impossible given p' for $P(q\,|\,p) = 0$ and 'q is almost certain given p' for $P(q\,|\,p) = 1$ when the corresponding entailments do not hold. This usage was introduced by I. J. Good. When it occurs a limiting process always seems to be involved. It must be distinguished from 'practically impossible' and 'practically certain', which we shall also use to indicate that probabilities are near 0 and 1, and usually could be made arbitrarily near if the number of observations was large enough.

2·3. The *principle of inverse probability* is an immediate consequence of the product rule. Let p be the initial data, θ a set of additional data, $q_1, q_2 \ldots q_n$ a set of hypotheses. Then

$$P(q_r.\theta\,|\,p) = P(q_r\,|\,p)\,P(\theta\,|\,q_r.p) \tag{1}$$

and also
$$= P(\theta\,|\,p)\,P(q_r\,|\,\theta.p) \tag{2}$$

and therefore the ratio

$$\frac{P(q_r\,|\,\theta.p)}{P(q_r\,|\,p)\,P(\theta\,|\,q_r.p)} = \frac{1}{P(\theta\,|\,p)} \tag{3}$$

is the same for all q_r. This is the principle of inverse probability. If the q_r are an exclusive and exhaustive set, the sum of the numerators of the expressions on the left is 1, and

$$P(q_r \mid \theta.p) = \frac{P(q_r \mid p)\, P(\theta \mid q_r.p)}{\Sigma P(q_r \mid p)\, P(\theta \mid q_r.p)}. \tag{4}$$

It is convenient to call $P(q_r \mid p)$ the prior probability† of q_r, $P(q_r \mid \theta.p)$ the posterior probability, and $P(\theta \mid q_r.p)$ the likelihood. Then the theorem can be stated

Posterior probability $\propto$ prior probability $\times$ likelihood.   (5)

The likelihood is the same as the 'direct probability' of many writers.

This theorem (due to Bayes)‡ is to the theory of probability what Pythagoras's theorem is to geometry. We can see its relation to scientific inference at once by considering a few special cases.

$P(q_r \mid \theta.p)$ can hardly ever be unity, since this would require that all the terms except one in the denominator of (4) are zero. But if θ has a small probability on all the hypotheses suggested except one, say q_1, and a large probability on that one, and the prior probabilities of the various q_r are comparable, then the posterior probability of q_1 may be nearly 1. This is the type of inference known as a *crucial test*.

Again, suppose that $P(\theta \mid q_1.p) = 1$, and that nevertheless $\sim\theta$ is verified. Then $P(q_1 \mid p. \sim\theta) = 0$. This explains how failure of a crucial test may lead a previously plausible hypothesis to be rejected.

It may happen that $P(\theta \mid q_r.p)$ is the same for all r. In that case the posterior probabilities are equal to the prior probabilities; in other words the new data do nothing to help us to decide between the hypotheses. This is the case of *irrelevance*.

If $P(q_r \mid p) = 0$, $P(q_r \mid \theta.p) = 0$. If a hypothesis is already known to be false, it will still be known to be false given any additional

† The prior probability is often called the *a priori* probability. This is an unfortunate usage because *a priori* is technically used in logic for propositions accepted independently of experience, and in probability theory would suggest that all prior probabilities were based only on the general datum H used above, whereas the prior probability is intended to express simply the probability at the start of an investigation and may have been influenced by many previous investigations. To make confusion worse, probabilities of throws of dice, which are likelihoods, are also often called *a priori* probabilities.

‡ *Phil. Trans.* **53**, 1763, 376–98.

evidence. This theorem adapts this deductive theorem to zero probabilities.

The principle of inverse probability therefore accounts for the use of the most striking type of experiment, the crucial test. We shall have many other applications, and a general theory of induction is impossible without it. Nevertheless it seems to fill many people with terror. Most statistical writers think it necessary to state that they do not propose to use inverse probability, and make an attack on the notion of prior probability. (After nearly a century the number is diminishing.) Some declare that prior probabilities cannot be assessed consistently, others that they are arbitrary, which means, if it means anything, that they can be consistently assessed in more ways than one. Some (whose logical ability in other matters is not negligible) make both criticisms.

We can state an answer to these criticisms. Prior probabilities can be assigned consistently in many ways; it is enough, and more than enough, that all considered propositions shall have positive probabilities on H. One of our problems is to find the most suitable selection for our purpose.

2·4. Bayes's original presentation followed different lines. He took as a fundamental idea 'expectation of benefit'. In deciding on a course of action, where the benefits may be attained by different courses but in each case depend on whether a problematic event happens, the possible failure of the event must be taken into account. For a consistent theory of behaviour it is supposed that courses of action can be arranged in an order according to the corresponding expectations of benefit. Let the possible benefit be a; the available evidence before action is taken is p. Suppose that a will be received if and only if an event q happens. Then denote the expectation of benefit by $E(a, q \mid p)$. This can be regarded as the value that it would be reasonable to stake on q. Bayes assumes that for given p and q $E(a, q \mid p)$ is proportional to a; then we can define the probability of q, given p, by
$$E(a, q \mid p) = aP(q \mid p). \tag{1}$$
Suppose now that q_1 and q_2 are two exclusive possibilities, and that a will be received if either happens. Then the expectation of benefit is the sum of those if a is to be received if q_1 happens, and if a is to be received if q_2 happens. Hence
$$E(a, q_1 \vee q_2 \mid p) = E(a, q_1 \mid p) + E(a, q_2 \mid p), \tag{2}$$

and on division by a we have the addition rule

$$P(q_1 \vee q_2 \mid p) = P(q_1 \mid p) + P(q_2 \mid p). \tag{3}$$

Again, suppose that a is to be received if and only if both q and r happen. Then

$$E(a, q.r \mid p) = aP(q.r \mid p). \tag{4}$$

We can proceed by testing q first and then r. Suppose that q has happened; then the available information is $p.q$. If r happens we should receive a. Hence at this stage the expectation of benefit is

$$E(a, r \mid p.q) = aP(r \mid p.q). \tag{5}$$

Return to the first stage. If q happens we do not receive a at once; what we get is the expectation of benefit conditional on r also happening, given $p.q$, that is, $E(a, r \mid p.q)$. If q does not happen we get nothing. Hence

$$E(a, q.r \mid p) = P(q \mid p) E(a, r \mid q.p), \tag{6}$$

whence

$$P(q.r \mid p) = P(q \mid p) P(r \mid q.p). \tag{7}$$

This is the product rule.

Thus the fundamental axioms of the theory arise as consequences of the hypothesis that a rational (economic) theory of action is possible. The principle of inverse probability, of course, follows as in 2·3.

Bayes's argument is developed in more detail by F. P. Ramsey.[†] I prefer myself to avoid the notion of additive degrees of expectation of benefit, which leads to paradoxes like the Petersburg problem, which depend on this notion and not on the nature of probability logic. For instance, the pleasures of two dinners on consecutive nights are certainly greater than that of two dinners on the same night, and some selection of benefits is needed if they are to be considered additive. Nevertheless the notion has been used in surprising circumstances. Eddington[‡] gave an account of the method of least squares, embodying a drastic attack on the principle of inverse probability; but in developing an account of the method he introduced comparisons of expectations of loss, and therefore had all the postulates needed to prove the principle by Bayes's method.[§] Wald, still more recently, has developed a theory of

[†] *Foundations of Mathematics* (1931).
[‡] *Proc. Phys. Soc.* **45**, 1933, 271–82.
[§] Jeffreys, *Phil. Mag.* (7) **22**, 1936, 337–59.

quality tests, also rejecting inverse probability but comparing expectations of benefit. Like Eddington, he was in a position to prove the principle from his postulates and only omitted to do so.

Until recently very little biographical information was available about Bayes, but Professor G. A. Barnard has traced his life.† He was born in 1702, became a Presbyterian minister at Tunbridge Wells, was elected to the Royal Society in 1742, retired from his ministry about 1750, and died in 1761.

2·5. There is a genuine difficulty in making use of the principle of inverse probability, which is not yet completely solved, but there is no reason to believe it insoluble, and there is some reason to suppose that it would be solved fairly quickly if as much ingenuity was applied to it as is applied to other branches of logic. The point is this. At the beginning of any actual experiment we have a certain amount of information, which we have denoted by p. This usually includes much previous observational information, and it can be asked: what were the probabilities before you had that information? The scientist may be able to produce some less detailed information, but the same question can be repeated, and he can be driven to face the question: what were the probabilities before you had any observational information at all? (If the expression '*a priori* probabilities' is used at all, it should be restricted to such probabilities. I prefer to call them *initial probabilities*.) When probability theory is regarded as a type of logic, this is seen to be a legitimate question. Every logical system must start somewhere, and the question simply amounts to 'where do we start?' Without any observational information our only possible information is the general principles of the theory. These must certainly include the axioms of pure mathematics and the addition and product rules of probability theory, but subject to these we could still achieve consistency in an infinite number of ways. Something else is certainly needed. We have, however, a valuable suggestion in a working rule given by Whitehead and Russell: when there is a choice of axioms, we should choose the one that enables the greatest number of inferences to be drawn. We are concerned with the probabilities of scientific laws, which are expressed by equations in mathematical form. Most of them

† *Biometrika* **45**, 1958, 293–315. The first three pages are biographical, the rest a reproduction of the 1763 paper.

contain what are called adjustable parameters. For instance, 'a mustard flower has six stamens' contains the number 6, and other flowers have different numbers of stamens. 6 in this law is determined by observation, but the form of the law at least supposes that the number of stamens is a positive whole number.

'The gravitational attraction between two bodies is proportional to mm'/r^2, where m, m' are the masses and r the distance between them', involves the two masses and expresses the fact that for given bodies (m, m' kept the same) but varying values of r, the ratio of the gravitational force to mm'/r^2 is a constant, f; the actual value of this constant is not known initially but has to be inferred from the observations. If we take different pairs of masses and can measure the masses otherwise, we find that f is also independent of m and m'. Here f is an adjustable parameter, but with other values of f the law would have the same form. There is here no question of f being required to be a whole number. Scientific laws can then be classified primarily according to their form. For a given form the laws may differ according to the values of adjustable parameters in them, but the values of the parameters cannot even be considered until the form is specified.

Whitehead's and Russell's principle suggests that laws should be classified primarily according to their form; if we adopt a certain initial probability for a law of a given form in one application, we should adopt the same value for it in other applications. The principle is not an empirical generalization and it would be absurd to ask for experimental evidence for it. Its purpose, in the present context, is simply to reduce arbitrariness.

Two theorems suggest how to apply this rule. Suppose that q is a hypothesis and that it entails a sequence of observable consequences $p_1, p_2, p_3 \ldots$. Denote the initial data by H. Suppose further that all consequences up to p_{n-1} are verified. Then by the product rule

$$P(p_1 . q \mid H) = P(q \mid H) P(p_1 \mid qH) = P(p_1 \mid H) P(q \mid p_1 . H). \quad (1)$$

Since q entails p_1, $P(p_1 \mid q . H) = 1$. Then

$$P(q \mid p_1 . H) = P(q \mid H)/P(p_1 \mid H). \quad (2)$$

Similarly (by substituting p_2 for p_1 and $H.p_1$ for H)

$$P(q \mid p_2.p_1.H) = \frac{P(q \mid p_1.H)}{P(p_2 \mid p_1.H)} = \frac{P(q \mid H)}{P(p_1 \mid H)P(p_2 \mid p_1.H)}. \quad (3)$$

and in general

$$P(q \mid p_1.p_2.\ldots.p_n.H)$$
$$= \frac{P(q \mid H)}{P(p_1 \mid H)P(p_2 \mid p_1.H)\ldots P(p_n \mid p_1.p_2\ldots p_{n-1}.H)}. \quad (4)$$

Now $P(q \mid H) \geqslant 0$, and all of $P(p_1 \mid H), \ldots P(p_n \mid p_1 \ldots p_{n-1}.H)$ are $\leqslant 1$. If $P(q \mid H) = 0$, the left side is always zero and q will never acquire a positive probability.

If $P(q \mid H) > 0$, the expressions on the right (as n increases) form a non-decreasing sequence of positive terms, which must tend either to infinity or to a finite limit as n tends to infinity. Suppose if possible that an infinite number of the factors $P(p_n \mid p_1 \ldots p_{n-1}.H)$ for different n were less than $\alpha < 1$. Then for sufficiently large n the denominator would be $< P(q \mid H)$, and $P(q \mid p\ldots.p_n.H) > 1$. This is impossible since this expression is a probability. Hence for any α only a finite number of factors can be $\leqslant \alpha$, and therefore, since α can be as near 1 as we like,

$$P(p_n \mid p_1.p_2.\ldots.p_{n-1}.H) \to 1. \quad (5)$$

We have therefore two alternatives. Either $P(q \mid H) = 0$, in which case the law has zero probability however often it is verified; or $P(q \mid H) > 0$, and then (5) follows. The first alternative would be an acceptance of failure (made by many philosophers who accept the theory up to the principle of inverse probability). The second leads to the conclusion that with increasing number of verifications the probability of success of the next test tends to 1, and therefore is in accordance with the belief that scientific inference can in suitable conditions be carried out with something near certainty. We may therefore accept the second alternative without hesitation. We note that the conclusion (5) does not contain q at all; it would still stand if q was false. It does not say that q ever acquires a high probability, merely that inferences from it do so. In the customary approach by way of causality, it would say that if q is false, the law that is true must be such as to have led, up to n trials, to the same inferences as q, and may be expected to lead to the same consequence at the next trial.

I do not adopt this approach because it begs the question of whether any general law is true at all, and does not avoid the word 'expected' or some equivalent, which implies the notion of probability; it therefore avoids no hypothesis and introduces an extra one. More is true, as Huzurbazar has pointed out.† We can write

$$u_n = P(p_1 p_2 \ldots p_n \,|\, H) = \frac{P(q\,|\,H)}{P(q\,|\,p_1\,p_2\ldots p_n H)}. \tag{1}$$

Then $u_n \geqslant P(q\,|\,H)$, a fixed positive number independent of n. Also

$$u_{n+1} = u_n P(p_{n+1}\,|\,p_1\ldots p_n H) \leqslant u_n. \tag{2}$$

Hence the sequence $\{u_n\}$ is non-increasing and tends to a limit

$$l \geqslant P(q\,|\,H) > 0. \tag{3}$$

Therefore for all n, m tending to infinity in any manner

$$\frac{u_{n+m}}{u_n} \to \frac{l}{l} = 1. \tag{4}$$

But this is $$P(p_{n+1} p_{n+2} \ldots p_{n+m}\,|\,p_1\ldots p_n H). \tag{5}$$

Hence with a sufficient number of verifications the probability that *all* future tests will give verifications tends to certainty.

H is understood in the above to be the general principles of the theory, which we are trying to state as precisely as is possible in the present state of analysis.

This leads at once to a further conclusion. The number of mutually exclusive scientific laws that might possibly be true, before we have any observational evidence, is presumably infinite. But the sum of their probabilities cannot exceed 1. Hence they form a finite or enumerable set (see Appendix I). At this point we have reached a principle that is not part of classical logic and must be included in H if a formal account of scientific procedure is to be possible.

For laws that concern only counting there is no difficulty in supposing that they form an enumerable set. But laws that contain adjustable parameters would apparently form a set with the number of the real numbers, at least if every value of a parameter is to be regarded as giving a separate law. We must therefore classify

† *Proc. Camb. Phil. Soc.* **51**, 1955, 761–2.

together laws that differ only in respect of the values of adjustable parameters. We have already had a suggestion that this should be done; but now we see that it must be done.

The question now arises whether the admissible laws, if restricted to form an enumerable set apart from adjustable parameters, would be numerous enough for scientific needs. The set considered at any moment is finite, so there is no practical difficulty. So far as there is a problem, it would concern future attempts at specifying a wide range of laws that might not admit classification into any of the types at present known to pure mathematics. However, the class of differential equations of finite order and degree, with rational coefficients, certainly includes all laws of classical physics; and they form an enumerable set. Further, their solutions bring in adjustable parameters. Thus we have been led to a condition that is in fact satisfied by classical physics. Further, the quantum theory modifies the classical differential equations in standard ways, and we can at least hope that any new ones will not be much more numerous than the classical ones. Without insisting on any particular set of laws as the only ones possible, therefore, we can state the principle in an acceptable form as follows. *The set of all possible forms of scientific laws is finite or enumerable, and their initial probabilities form the terms of a convergent series of sum* 1. We shall call this principle the *simplicity postulate*, for the following reason.

Scientific practice is possible only by considering laws in an order. We cannot examine all possible laws at once; for one thing, nobody has thought of them. What we ordinarily do is to adopt some law that fits the observations within a range of what we call error. If we find at some time that the observations cannot be fitted by a law of this form within the range of error that we believe possible we modify the form of the law. The modified law is then taken as standard until further modification is needed. In practice the modifications usually imply increasing complexity. Occasionally they do not, and then the behaviour of scientists is of special interest. Einstein's general theory of relativity was simpler than any previous theory that tried to cover the phenomena of light and gravitation simultaneously. The modern quantum theory, started by Heisenberg, was also simpler than any previous theory that covered the same data. In both cases the new theories made new predictions, which were verified and in any case would be regarded as evidence for the laws proposed. But in addition to this the very fact that the new laws were simpler than the old ones was widely and influentially claimed to be a reason for accepting them. This amounts to saying that in the absence of observational evidence, the simpler law is the more probable and the initial probabilities can be placed in an order. To identify this with our postulate, all that we have to say is that the order of decreasing initial probabilities is that of increasing complexity.†

2·6. The use of the notion of error in the last paragraph brings us to a further theorem. Suppose that on a hypothesis q the value of a measurement x is predicted to be within a range $\alpha \leqslant x < \alpha + \epsilon$. θ is the result that the measurement is actually found in a range $x_0 < x < x_0 + \delta x$, with $\alpha < x_0 < x_0 + \delta x < \alpha + \epsilon$. Suppose that on $\sim q$, x might be anywhere within a range of length E, much greater than ϵ, and including α to $\alpha + \epsilon$. Then if the initial data are p,

$$P(\theta \mid q.p) = k\delta x/\epsilon; \quad P(\theta \mid \sim q.p) = l\delta x/E \tag{1}$$

where k, l are of the order of unity; and

$$\frac{P(q \mid \theta.p)}{P(\sim q \mid \theta.p)} = \frac{k}{l}\frac{P(q \mid p)}{P(\sim q \mid p)}\frac{\delta x/\epsilon}{\delta x/E} = \frac{k}{l}\frac{E}{\epsilon}\frac{P(q \mid p)}{P(\sim q \mid p)}. \tag{2}$$

Thus the more precise the inferences given by a law are, the more its probability is increased by a verification, even if the contradictory law also gives a prediction consistent with the observation. A single verification of a precise prediction may send up the probability of the law nearly to 1. If $P(q \mid p)/P(\sim q \mid p)$ is not a small fraction, it follows that $P(q \mid \theta.p)/P(\sim q \mid \theta.p)$ is large, on account of the large factor E/ϵ.

We may say that to make predictions with great accuracy increases the probability that they will be found wrong, but in compensation they tell us much more if they are found right. We are thus led to a solution of the problem of 'scientific caution'. Everybody agrees on the need for caution. Some say that it is 'speculative'

† Wrinch and I (*Phil. Mag.* **42**, 1921, 369–90), in reaching the simplicity postulate in this form, did not use the theorem of Appendix I. My knowledge of the theory of sets at the time did not include this theorem, and my partner may have thought it too obvious to need stating.

to try to draw inferences far from the original data (*speculative* being a term of abuse). Some say that no inferences at all should be drawn unless we are in a position to verify them directly. On the other hand the further we get from the original data, the more informative a verification is. The best procedure, accordingly, is to state our laws as precisely as we can, while keeping a watch for any circumstances that may make it possible to test them more strictly than has been done hitherto.

It may happen that over a certain range of circumstances several laws make the same predictions. In that case the verification of the inferences affects all their probabilities in the same ratio; we may say that the observations are irrelevant as between the hypotheses. If the hypotheses would make the same predictions in all circumstances, we shall never be able to decide between them; we must regard them, not as different laws, but as different ways of stating the same law.

The simplicity postulate is therefore in accordance with the more striking features of scientific method. It could not be disproved by observation, since the number of laws considered up to any moment of time is finite; hence it is not an empirical hypothesis and must be taken as belonging to the general principles. We have, however, shown only that it accounts for the main features. To get numerical answers in actual applications we must try to state it more precisely. This involves deciding on the order of decreasing initial probabilities and then assigning actual numbers. There is a choice at the start, because whatever order is chosen, if the probability of the nth law in the series is a_n, the conditions would be satisfied if

$$a_n = 2^{-n}$$

or if
$$a_n = 6/\pi^2 n^2,$$

or by many other assignments. This is a matter of minor importance because, as we shall see, any plausible assignment would lead to nearly the same results in most actual applications. What is more serious is that we have to state *what* order is the order of decreasing initial probability. Simplicity gives us a hint, but only a hint, because we should need a quantitative definition of simplicity itself. Many writers state that no such definition is possible, and regard the notion of simplicity as incurably vague. However, a definition is certainly possible. A law may contain quantities that by definition must be integers; quantitative laws in general contain also quantities that are necessarily variable, but also adjustable parameters that can have any values within certain ranges. Then we could define the complexity of a law as the sum of the absolute values of the integers in it, together with the number of adjustable parameters. If this sum is different for different ways of stating the law, take the least value. Thus it is possible to give a precise definition of the complexity of any law that is expressible in finite terms. Where several laws have the same complexity, the probability available for laws of that complexity can be shared equally between them. Hence it is possible to assign numbers to represent the initial probabilities of laws in a specifiable way, and in fact in an infinite number of specifiable ways.

Does that condemn the whole system as arbitrary? No, for two reasons. First, once the problem is clearly stated and recognized not to have a unique answer, it could be referred to an international body of scientists, who could recommend one of the alternatives for general use. The decision might be held to represent an expression of average human prejudice, but, after all, human prejudices exist and such a decision would bring them into the open. At present a person may have quite different standards for his own hypotheses and other people's, without ever feeling the need to mention the difference. There are people who appear to attach probability 1 to their own views and others who attach probability $\frac{1}{2}$ to any hypothesis, however often it has been verified; and both types would maintain that they are scientific. The reason why they continue to exist is the lack of attention to the analysis of scientific method.

Secondly, we shall find that in cases where there are many relevant data or where a crucial test is possible, the posterior probabilities are affected very little even by quite considerable changes in the prior probabilities; a course of action would be affected by such changes only in circumstances where the ordinary scientific comment would be 'get some more observations'. We can make considerable progress while still leaving some latitude in the choice of initial probabilities. This book is mainly concerned with problems where this is possible.

In (2) it is assumed that δx is not zero. If $\delta x = 0$ we could not cancel it, and the argument would fail. But if $\delta x = 0$ we should (for continuous distributions) have $P(\theta \mid q.p) = P(\theta \mid \sim q.p) = 0$. We should then be considering zero probabilities, and in 2·2 we have had a warning that this is dangerous. It looks as if exact observations would make the theory unworkable. However, all measures have a range of uncertainty, and this is not important.

CHAPTER III

SAMPLING

Trifles, light as air, shall waft a belief into the soul, and plant it so immovably within it, that Euclid's demonstrations, could they be brought to batter it in breach, should not all have power to overthrow it.

LAURENCE STERNE, *Tristram Shandy*

3·0. We are now in a position to discuss the application of the methods of probability theory to the problem of sampling. This was the first problem treated by means of the principle of inverse probability by Bayes and Laplace. Modern criticism has shown that their treatment needs some modifications, but nevertheless it needs attention both because in a wide class of problems it is very accurate and because its very failure to deal adequately with some other problems gave the most direct clue to the kind of revision that has turned out to be necessary.

We need first a general theorem on what is called *chance*. Suppose that on a hypothesis q an event p may happen with probability α. We have

$$P(p \mid q) = \alpha; \quad P(\sim p \mid q) = 1 - \alpha. \tag{1}$$

Suppose that we make a series of trials; for each trial the results of previous trials are among the data, and hence the probability may not be the same. Thus we should write

$$P(p_n \mid q p_1 p_2 \dots p_{n-1}) = \alpha_n. \tag{2}$$

At this point we can omit the dots in writing logical products in cases where there is no risk of ambiguity.

Now as a special case suppose that the results of previous trials do not affect α_n; that is, α_n is always equal to α, however many of the p_r in the data are replaced by not-p_r. Then the probability that p will happen at the first and second trials is α^2; that it will happen at the first and fail at the second is $\alpha(1 - \alpha)$; that it will fail at the first and happen at the second is $(1 - \alpha)\alpha$; and that it will happen neither time is $(1 - \alpha)^2$. It can then be shown by mathematical induction that the probability that in n trials it will happen l times and not happen $n - l$ times, in any specified order, is

$\alpha^l(1-\alpha)^{n-l}$. Further, since all of these arrangements are mutually exclusive, the probability that p will happen l times and fail $n-l$ times, for all possible orders, is the sum over all possible orders. The number of possible orders is the number of ways of selecting l places out of n, which is

$$^nC_l = \frac{n!}{l!\,(n-l)!}.$$ (3)

(I use the old-fashioned notation ^nC_l instead of the more modern $\binom{n}{l}$ because it can be printed on one line.) Then the probability that p will happen l times out of n, all orders being taken together, is

$$^nC_l\,\alpha^l(1-\alpha)^{n-l}.$$ (4)

If we write $1-\alpha=\beta$, this is the general term in the expansion of $(\alpha+\beta)^n$ by the binomial theorem, and hence the rule is known as the binomial law. We note that it depends on q, which must be taken to specify the value of α and the condition of irrelevance of previous trials, and also on n. Thus q and n are the data. The proposition 'p will happen l times' will be abbreviated simply to l. Then we write the result as

$$P(l\,|\,qn) = \frac{n!}{l!\,(n-l)!}\alpha^l(1-\alpha)^{n-l}.$$ (5)

It is easy to show that this is greatest for $l=l_0$, where

$$(n+1)\alpha-1 < l_0 < (n+1)\alpha.$$ (6)

l_0 never differs from $n\alpha$ by more than 1. What is more interesting is that the expectation of l is always exactly $n\alpha$. By the expectation of l we mean the sum $\Sigma lP(l\,|\,qn)$ over all possible values of l; we write this as $E(l\,|\,qn)$. Then

$$E(l\,|\,qn) = \sum_{l=1}^{n} \frac{n!}{(l-1)!\,(n-l)!}\alpha^l(1-\alpha)^{n-l}$$

$$= n\alpha \sum_{l=1}^{n} \frac{(n-1)!}{(l-1)!\,(n-l)!}\alpha^{l-1}(1-\alpha)^{n-l}$$

$$= n\alpha\{\alpha+(1-\alpha)\}^{n-1} = n\alpha.$$ (7)

Also $\qquad (l-n\alpha)^2 = l(l-1) - (2n\alpha-1)\,l + n^2\alpha^2$ (8)

and

$$E\{l(l-1)\,|\,qn\} = \sum_{l=2}^{n} \frac{n(n-1)(n-2)!}{(l-2)!(n-l)!}\,\alpha^l(1-\alpha)^{n-l} = n(n-1)\alpha^2 \quad (9)$$

whence $\quad E\{(l-n\alpha)^2\,|\,qn\} = n(n-1)\alpha^2 - (2n\alpha - 1)\,n\alpha + n^2\alpha^2 \quad (10)$

$$= n\alpha(1-\alpha). \quad (11)$$

$E(l\,|\,qn)$ is usually called the mean of l for the law, and

$$E\{(l-n\alpha)^2\,|\,qn\}$$

the second moment or the variance. $E(l\,|\,qn)$ is also often called the expected value, which is a most unfortunate expression because in a sense any integral value of l from o to n is expected, since it is possible; and $n\alpha$ may not be an integer and then *no* possible value of l is equal to $n\alpha$. The interesting feature of the second moment is that, though $(l-n\alpha)^2$ can range from o to the greater of $n^2\alpha^2$ and $n^2(1-\alpha)^2$, its expectation is never greater than $\frac{1}{4}n$ whatever α may be. The probabilities of l are small except within a range of l about $n\alpha$ with a length of order $\sqrt{n}$. This can be expressed in a convenient way when n is large and α is not equal to o or 1, namely†

$$P(l\,|\,qn) = \{2\pi n\alpha(1-\alpha)\}^{-\frac{1}{2}} \exp\left\{-\frac{(l-n\alpha)^2}{2n\alpha(1-\alpha)}\right\}(1 + O(n^{-\frac{1}{2}})). \quad (12)$$

This looks complicated; but for large n the factorials are enormous, whereas this expression is easy to calculate. If we take a quantity δl large compared with 1 and small compared with $\{n\alpha(1-\alpha)\}^{\frac{1}{2}}$, which will be possible if α is not o or 1 and n is large enough, the number of values of l in a range δl is δl, and in such a range the exponent does not vary much. Then if we put

$$n\alpha = \lambda, \quad n\alpha(1-\alpha) = \sigma^2, \quad l - \lambda = x,$$

the probability that l lies in a range δl is nearly

$$\frac{1}{\sqrt{(2\pi)}\,\sigma} \exp\left(-\frac{1}{2}\frac{x^2}{\sigma^2}\right)\delta l. \quad (13)$$

This probability distribution is of such frequent occurrence that it is known as the *normal law*. It is approximately satisfied by errors of measurement. It is often called the Gaussian law of error, but

† Proof in my *Theory of Probability* (1950).

Laplace had applied it to errors of observation before Gauss, and the present application was given still earlier by de Moivre (English, in spite of his name). Karl Pearson, to reduce international complications, recommended the use of 'normal law'. In the form above σ is called the standard error. The behaviour of the function is striking. We write $u = x/\sigma$,

$$\phi(u) = \int_{-u}^{u} \frac{1}{\sqrt{(2\pi)}} \exp\left(-\tfrac{1}{2}u^2\right) du \tag{14}$$

and have the following skeleton table:

u	$\exp\left(-\tfrac{1}{2}u^2\right)$	$\phi(u)$
0	1·0000	0·0000
1	0·6065	0·6827
2	0·1353	0·9545
3	0·0111	0·9973
4	0·00034	0·9999

Notice the rapid decrease of the exponential for $u > 2$. The probability of a deviation greater than 2σ (both sides being taken together) is about 0·05, and of one greater than 3σ is 0·0027. We may say that values of x/σ outside the range -3 to 3 are practically impossible. Other interesting properties of the normal law are:

The probability of a deviation of either sign less than the standard error is roughly $\tfrac{2}{3}$. This is convenient to remember.

The second derivative of the exponential vanishes at $x = \pm\sigma$. These points are at the inflexions on a graph.

The expectation of x is 0 and that of x^2 is σ^2. In terms of l, the expectation of l is $n\alpha$ and that of $(l - n\alpha)^2$ is $n\alpha(1-\alpha)$. Thus the approximation reproduces the mean and the second moment of the binomial law exactly.

In such a case as this, where, given certain parameters, the probability of an event is the same at every trial, no matter what may have happened at previous trials, we say that the probability is a *chance*; the term was used in this sense by N. R. Campbell and revived by M. S. Bartlett.

3·1. If we take a fixed positive number ϵ, however small, the probability that $l - n\alpha$ lies between $\pm n\epsilon$, and therefore x/σ between $\pm n\epsilon/\sigma$, is nearly $\phi(n\epsilon/\sigma)$. But n/σ tends to infinity like $n^{\frac{1}{2}}$ as n increases. Hence the probability tends to 1. Hence whatever ϵ may be it is practically certain that for a sufficiently large sample the ratio l/n will be between $\alpha \pm \epsilon$. This theorem was given by James Bernoulli and is called the 'law of large numbers' or the 'law of averages'. In this form it does *not* prove that as n tends to infinity l/n tends to α. What can be proved, with more difficulty, is that there is probability 1 that l/n tends to α. But it is not logically certain that l/n will tend to α. This is easily seen. It is essential to the theorem that α is not equal to 0 or 1. Hence, whatever the results of previous trials, either result is always possible at the next. Thus, if we write 1 for each occasion when the event happens, and 0 for when it fails, any of the following series is possible:

$$000000000\ldots.$$
$$111111111\ldots.$$
$$010101010\ldots.$$
$$011011011\ldots.$$
$$101100001111111110\ldots.$$

In the last series the number in each block is the total number of previous trials. But the ratios l/n tend respectively to 0, 1, $\frac{1}{2}$, $\frac{2}{3}$, and to no limit whatever—in the last case l/n oscillates between $\frac{1}{3}$ and $\frac{2}{3}$. The data do not entail that l/n tends to α. We shall return to this point later, as it is crucial in the attempt to define a class of probabilities by a limiting process (p. 193).

3·2. *Poisson's rule.* An interesting case of the binomial law is where $n\alpha$ is given, equal to r say, but n is large. In this case, if l is not too large,

$$^nC_l\,\alpha^l(1-\alpha)^{n-l} = \frac{n(n-1)\ldots\ldots(n-l+1)}{l!\,n^l}\,r^l\left(1-\frac{r}{n}\right)^{n-l}. \qquad (1)$$

With r and l fixed and n tending to infinity this tends to $r^l e^{-r}/l!$. This is Poisson's rule. The probability that the event will happen l times is e^{-r} times the term in r in the expansion of e^r. It is applicable when we are considering a very small chance and compensate for it by having a very large number of trials, and we want to know the probability that any particular number of them will happen. The classical example was given by von Bortkiewicz in a study of the number of soldiers in certain Prussian army corps that were killed in a year by the kicks of horses. The number of soldiers was large; the chance for an individual was small, but the product was appreciable. One of great physical importance was in a study by

Rutherford and Geiger of the numbers of α-particles emitted by a specimen of a radioactive substance in successive intervals of $\frac{1}{8}$ minute. The number of radioactive atoms was very large, and the chance of any one breaking up in a particular interval very small; the product was about 4.

3·3. The binomial rule can be extended. Suppose that at each stage p alternatives are possible; that the chance of the rth alternative occurring is α_r (so that $\Sigma\alpha_r = 1$); and that n trials are made. The chance that the alternatives will occur $n_1, n_2, \ldots n_p$, times is

$$\frac{n!}{n_1! n_2! \ldots n_p!} \alpha_1^{n_1} \alpha_2^{n_2} \ldots \alpha_p^{n_p}. \tag{1}$$

The expectation of n_r is $m_r = n\alpha_r$. This is the multinomial rule. Pearson showed that for variations of n_r, if the m_r are large, the probability that the n_r will have a given set of values is approximately proportional to $\exp(-\frac{1}{2}\chi^2)$, where

$$\chi^2 = \Sigma \frac{(n_r - m_r)^2}{m_r}. \tag{2}$$

It can be verified that for $n = 2$, $-\frac{1}{2}\chi^2$ reduces to the exponent in the normal approximation to the binomial law. A particular value of χ^2 could arise from many different combinations of values of the n_r. Pearson went further and showed that the total probability of χ^2 falling in a particular range $\delta\chi^2$ was proportional to

$$\chi^{p-2} \exp(-\frac{1}{2}\chi^2) \delta\chi, \tag{3}$$

subject to approximations similar to those used in getting the normal law as an approximation to the binomial.

χ^2 has many applications. For one thing, if the α_r are unknown and the n_r are observed, the likelihood is nearly proportional to $\exp(-\frac{1}{2}\chi^2)$, and consequently χ^2, regarded as a function of the α_r, sums up all the information that the observations can give about the α_r. For another, it can arise in a still more general way. Suppose that n measurable quantities x_r have normal probability distributions, centred on λ_r, each with standard error σ_r. We define now

$$\chi^2 = \Sigma \frac{(x_r - \lambda_r)^2}{\sigma_r^2}. \tag{4}$$

This reduces to (2) if we take $x_r = n_r$, $\lambda_r = m_r$, $\sigma_r = \sqrt{m_r}$.

It can be shown that the probability that χ^2 will lie in a particular range $\delta\chi^2$ is proportional to

$$\chi^{n-1}\exp\left(-\tfrac{1}{2}\chi^2\right)\delta\chi. \tag{5}$$

Much more is true. Suppose that the λ_r are given functions of m quantities μ_s, and that we try to make χ^2 a minimum by making a suitable choice of the μ_s, say m_s. Let l_r be the value of λ_r corresponding to $\mu_s = m_s$, and calculate χ^2 for $\lambda_r = l_r$. The probability distribution of *this* χ^2 is now

$$\chi^{n-m-1}\exp\left(-\tfrac{1}{2}\chi^2\right)\delta\chi. \tag{6}$$

Every parameter estimated in this way reduces the index by 1. The reason why the index in (3) is $p-2$ instead of $p-1$ is that, given p and n, only $p-1$ of the n_r can be assigned independently, since $\Sigma n_r = n$. The difference $n-m$ is called the number of *degrees of freedom* and usually denoted by ν. To see in another way how it arises, suppose that we have n observed data and that there are m parameters. To minimize χ^2 needs m equations to give estimates of the parameters. Then $n-m=\nu$ is the number of ways of altering the data independently subject to the restriction that the estimates remain unaltered.

We often want to estimate the parameters in a law, and in many cases there are easier ways of doing it than by minimizing χ^2. But the difference in the results is always small, and χ^2 remains useful as a general check on the correctness of the hypotheses. The expectation of χ^2 is always ν. The probability distribution of χ^2 has been calculated in detail for different values of ν, but as a general rule it can be stated that χ^2 has a chance of about $\tfrac{2}{3}$ of lying between $\nu \pm \sqrt{(2\nu)}$ and about a 5 per cent chance of lying outside $\nu \pm 2\sqrt{(2\nu)}$. So, if after estimation we compute χ^2 and find that it lies between $\nu \pm \sqrt{(2\nu)}$, we can say that this is what we should have expected. If it lies in the range $\nu + \sqrt{(2\nu)}$ to $\nu + 2\sqrt{(2\nu)}$ it is no more than we should expect to occur fairly often. If, however, χ^2 exceeds $\nu + 2\sqrt{(2\nu)}$ an event improbable on the original hypothesis has occurred, and it affords ground for examining the correctness of the hypothesis more closely. In this usage the χ^2 distribution is regarded as a *significance test*. It does not give the probability of the hypothesis; what it does give is a convenient, though rough, criterion of whether closer investigation is needed.

As an example, we take the data given by Rutherford and Geiger for their counts of α-particles. The first column gives the number of α-particles in an interval, the second the number of intervals when that number were ejected. The total number of intervals was 2608, the total number of particles 10097, and the Poisson parameter is taken to be the ratio 3·87. The expectation of the number of intervals when m particles would be ejected, according to the Poisson rule, is $2608 e^{-3\cdot87} 3\cdot87^m/m!$. The comparison is then as follows:

No.	Obs.	Exp.	$O-E$	χ^2
0	57	54	+ 3	0·2
1	203	211	− 8	0·3
2	383	407	−24	1·4
3	525	525	0	0·0
4	532	508	+24	1·1
5	408	393	+15	0·6
6	273	254	+19	1·4
7	139	140	− 1	0·0
8	45	68	−23	7·8
9	27	29	− 2	0·1
10	10	11	− 1	0·1
11	4	4 ⎫		
12	0	1 ⎪ +1		0·2
13	1	0 ⎬		
14	1	0 ⎭		

$$\text{Total } \chi^2 = \Sigma \frac{(O-E)^2}{E} = 13\cdot3.$$

The numbers for 11 and more observations have been taken together because the approximations used in deriving the χ^2 rule become poor when the expectations are less than about 5. Thus we have 12 numerical data. The expectations have been made to fit the total number of intervals and an estimate of the Poisson parameter, so that the estimate is on 10 degrees of freedom. In this case χ^2 would be expected (with probability about $\tfrac{2}{3}$) to lie between $10 \pm \sqrt{(20)}$, or between 5·5 and 14·5, and it does. This is enough to say that the data are in good accordance with the Poisson rule.

The large contribution 7·8 from $m=8$ might suggest a discrepancy, if taken by itself, but out of 12 data it is not surprising that the largest should have such a value.

Without χ^2 to give some quantitative standard of agreement, a person might inspect the $O-E$ values and say 'most of the discrepancies are small compared with the whole variation, so agreement is very good'. Another might say 'the differences are surprisingly large; the law doesn't fit very well'. The essential point is that χ^2 gives an idea of *how* large discrepancies may arise by chance; without such a criterion people's guesses on this point may be wildly too large or too small.

As a contrast, we take the following data from Miss E. M. Newbold, on the numbers of accidents experienced by the employees in a factory during a year:

No.	Obs.	Calc.	χ^2
0	54	7·0	317
1	56	27·8	29
2	57	55·5	0
3	53	73·7	6
4	29	73·5	28
5	28	58·6	16
6	19	38·9	10
7	20	21·2	0
8	9	11·0	4
9	11	4·9	
10	11	2·0	
11	7	0·8	
12	10	0·4	
13	3	0·3	
14	3	0·1	
15	1	—	
16	2	—	
17	2	—	
18	0	—	
19	0	—	
20	1	—	
⩾9	51	8·5	213

χ^2 is 623 on $10 - 2 = 8$ degrees of freedom! This is an extreme case, and the approximations used in deriving the χ^2 rule are invalid (these assume in particular $|n_r - m_r| < m_r$); but it is clear that the discrepancies are far greater than could be expected to occur by chance if the Poisson rule was applicable. The explanation is that some people are much more liable to have accidents than others; the careful people give the excess at 0 and 1 accident, the careless give the long tail at high numbers. In this case a different law, known as the negative binomial, was found to give a good fit. The data of von Bortkiewicz did not show a similar discrepancy, mainly because nobody can be killed twice by the kick of a horse.

3·4. A slightly more general problem is that of random selection from a finite class. Let the class have number n and contain r members with a property ϕ. A set of m members is selected at random. This means that all selections of m are equally probable. The whole number of possible selections of m out of n is ${}^nC_m$. Consider the selections that contain l members with the property ϕ and $m-l$ with $\sim \phi$. These are ${}^rC_l{}^{n-r}C_{m-l}$ in number. Hence the probability that the sample will contain l ϕs is

$$P(l\,|\,qnrm) = \frac{{}^rC_l{}^{n-r}C_{m-l}}{{}^nC_m}. \tag{1}$$

It is easy to show that this has its greatest value if l is the integer nearest to rm/n; that is, the most probable sample is a fair sample.

Suppose that n, m, l are given and that the problem is to estimate r. Let the probability, given q, n, that there are precisely r in the original class be $f(r)$, and suppose that m is irrelevant to it. (The last condition is not trivial; for if r/n was very small we might persist in increasing m until we had found a moderate number of ϕs. This is analogous to the conditions that lead to the Poisson law.) Then

$$P(r\,|\,qnm) = f(r), \tag{2}$$

$$P(r\,|\,qnml) = \frac{f(r)\,{}^rC_l{}^{n-r}C_{m-l}}{\sum\limits_{r=0}^{n} f(r)\,{}^rC_l{}^{n-r}C_{m-l}}, \tag{3}$$

since ${}^nC_m$ does not depend on r. We can make no further progress without some knowledge of $f(r)$. If there is no previous reason to suppose one value of r more likely than another, $f(r)$ is the same for all values of r. Then

$$P(r\,|\,qnml) = \frac{{}^rC_l{}^{n-r}C_{m-l}}{\sum {}^rC_l{}^{n-r}C_{m-l}} = \frac{{}^rC_l{}^{n-r}C_{m-l}}{{}^{n+1}C_{m+1}}. \tag{4}$$

This is greatest when r/l is as near as possible to n/m; that is, the most probable value of r is got by supposing the sample to be a fair one. (For the method of summation, see Appendix II.)

Suppose we want the probability that the next trial will give a ϕ. Denote this proposition by p. Then

$$P(p\,|\,qnml) = \sum_r P(pr\,|\,qnml)$$

$$= \sum_r P(r\,|\,qnml)\,P(p\,|\,rqnml). \tag{5}$$

Given r, n, m and l there are $n-m$ things left in the class and $r-l$ of them have the property ϕ; then

$$P(p \mid rqnml) = \frac{r-l}{n-m},$$

and

$$P(p \mid qnml) = \sum_r \frac{{}^rC_l \, {}^{n-r}C_{m-l}}{{}^{n+1}C_{m+1}} \frac{r-l}{n-m}$$

$$= \sum \frac{r!}{l!(r-l-1)!} {}^{n-r}C_{m-l} \bigg/ \frac{(n+1)!}{(m+1)!(n-m-1)!}$$

$$= \frac{l+1}{m+2} \sum \frac{{}^rC_{l+1} \, {}^{n-r}C_{m-l}}{{}^{n+1}C_{m+2}}$$

$$= \frac{l+1}{m+2}. \tag{6}$$

Thus the probability that the next member will be a ϕ is independent of n; it depends wholly on the sample. If $l=m$ and m is large, this probability approaches 1 but is never equal to it.

This argument was long held to give a general explanation of induction. It is, however, inadequate, as we shall see in a moment. By an extension due to Pearson, if all m have been ϕ and we take a further sample of number m', the probability that these m' will also be ϕ is $(m+1)/(m+m'+1)$; thus there is probability $\frac{1}{2}$ that the next $m+1$ will be ϕ (see Appendix II). If $m'=n-m$, the result is

$$P(r=n \mid qmn, l=m) = \frac{m+1}{n+1}. \tag{7}$$

This can also be obtained directly from (4). Thus there would never be a high probability that a property is common to all members of a class unless nearly the whole of the class has been examined. This shows that the analysis of sampling procedure given so far is quite inadequate to account for the high probability that we often attach to a general law.

The result evidently depends on the form taken for $f(r)$, which we have taken constant. If $f(r) = \frac{1}{2}$ for $r=0$ and $r=n$, and is zero for all other values, a sample of one member would suffice to show that $r=0$ or n. On the other hand, suppose that r was known to be s, where $s \neq 0$ or n; then the posterior probability that $r=s$ is always 1. Then

$$P(p \mid qn, r=s, m, l) = \frac{s-l}{n-m} \tag{8}$$

and *decreases* as l increases. We should of course expect this; the more ϕ's there are in the sample the less likely we are to find one at the next trial if the total number of ϕ's is known. This is qualitatively different from the result of the Laplace theory, which indicates that the probability of a ϕ at the next trial increases with the number of ϕ in the sample. The existence of these cases is no argument against the theory, because we may quite well have previous information that some values of r are more likely than others, and $f(r)$ makes allowance for it possible. $f(r)$ can be taken constant *only* if we have no previous information about the value of r (as Bayes said again and again).

But $f(r)$ constant does not apply to ordinary scientific procedure. A single specimen of a plant not belonging to any known species would create a strong belief that it would have, like most plants, a set of inheritable characters that would always distinguish it from other species until the characters themselves become altered in the process of evolution. This situation can be covered by a different form for $f(r)$. It is simplified if we suppose n very large, with $r/n = \alpha$; in that case $3\cdot4\,(1)$ approximates to $3\cdot0\,(4)$.

3·5. *The significance problem for chances.* Suppose that there is a non-zero probability k that α has a particular value α_0; denote the proposition $\alpha = \alpha_0$ by q, which we shall call the *null hypothesis*. On the *alternative hypothesis q'* suppose that the probability of α is uniformly distributed between 0 and 1. Then

$$P(q \mid mH) = k; \quad P(l \mid mqH) = {}^mC_l\,\alpha_0^l(1 - \alpha_0)^{m-l}, \tag{1}$$

$$P(q'd\alpha \mid mH) = (1 - k)\,d\alpha; \quad P(l \mid mq'\alpha H) = {}^mC_l\,\alpha^l(1 - \alpha)^{m-l}, \tag{2}$$

whence $$P(q \mid mlH) \propto k\alpha_0^l(1 - \alpha_0)^{m-l}, \tag{3}$$

$$P(q'd\alpha \mid mlH) \propto (1 - k)\,\alpha^l(1 - \alpha)^{m-l}\,d\alpha, \tag{4}$$

the same factor being understood under both signs of proportionality. Then

$$\frac{P(q \mid mlH)}{P(q' \mid mlH)} = \frac{k}{1-k}\,\frac{\alpha_0^l(1 - \alpha_0)^{m-l}}{\displaystyle\int_0^1 \alpha^l(1 - \alpha)^{m-l}\,d\alpha} = \frac{k}{1-k}\,\frac{(m+1)!}{l!\,(m-l)!}\,\alpha_0^l(1 - \alpha_0)^{m-l}. \tag{5}$$

Suppose $\alpha_0 = 0$ and $k = \frac{1}{2}$; this says that there is a suggestion, as

likely as not to be right, that the chance of the event is zero. Then
(5) is zero unless $l=0$. If $l=0$ the ratio is

$$\frac{k}{1-k}(m+1)=m+1. \tag{6}$$

Thus on m trials with the event never occurring there will be odds
of $m+1$ to 1 that the chance is zero; similarly if the event occurs
every time there will be the same odds that the chance is 1. This is
in accordance with the principle that a high probability can be
attached to a general law by a moderate amount of evidence.

Now take $\alpha_0=\tfrac{1}{2}$, $k=\tfrac{1}{2}$, so that we are testing whether a chance is
even. Then we define

$$K=\frac{P(q\,|\,mlH)}{P(q'\,|\,mlH)}\bigg/\frac{P(q\,|\,H)}{P(q'\,|\,H)}. \tag{7}$$

and for a few small values of m and l we get

m	l	K	m	l	K	m	l	K
1	0	1	2	1	$\frac{3}{2}$	4	2	$\frac{15}{8}$
2	0	$\frac{3}{4}$	3	1	$\frac{3}{2}$	6	3	$\frac{35}{16}$
3	0	$\frac{1}{2}$	4	1	$\frac{5}{4}$	8	4	$\frac{315}{128}$
4	0	$\frac{5}{16}$	5	1	$\frac{15}{16}$	10	5	$\frac{693}{250}$
5	0	$\frac{3}{16}$	6	1	$\frac{21}{32}$			
7	0	$\frac{1}{16}$						

K first becomes less than $0\cdot1$ for $m=7$ and $l=0$ or 7, and first
becomes greater than 10 if $m=2l=160$. It takes a fairly large sample
for $\alpha_0=\tfrac{1}{2}$ to establish strong evidence for or against the null hypothesis, but it can be done.

This analysis covers such problems as those of genetics, where
there is a strongly suggested null hypothesis that a chance is $\tfrac{1}{2}$, $\tfrac{1}{4}$ or
some other rational fraction with a small integer as denominator.
It can also be used to test whether dice or coins are biased. It is our
first example of a significance test, in which the question asked is
whether a suggested value of a parameter is correct.

If K is small, so that the null hypothesis has a small probability,
we shall want an estimate of α on the alternative hypothesis. This is
given by

$$P(d\alpha\,|\,q'mlH)\propto \alpha^l(1-\alpha)^{m-l}d\alpha. \tag{8}$$

The coefficient of $d\alpha$ is greatest when $\alpha=l/m$; and for values near
l/m we have, nearly,

$$P(d\alpha\,|\,q'mlH)=\left\{\frac{m^3}{2\pi l(m-l)}\right\}^{\frac{1}{2}}\exp\left\{-\frac{1}{2}\frac{m^3}{l(m-l)}\left(\alpha-\frac{l}{m}\right)^2\right\}d\alpha. \tag{9}$$

Thus the posterior probability of α nearly follows the normal law with standard error $\{l(m-l)/m^3\}^{\frac{1}{2}}$. It is worth while to notice that this is quite considerable for moderate m. Suppose $m = 100$, $l = 50$. Then

$$\frac{l(m-l)}{m^3} = \frac{50.50}{100^3} = \frac{1}{400}; \quad \sigma = 0.05.$$

There is a probability of about $\frac{2}{3}$ that the chance lies between 0.45 and 0.55. To determine a chance of 0.5 with a standard error of 0.01 would need a sample of 2500, and even then the probability that the estimate was wrong by over 0.02 would be about 5 per cent. Popular ideas on the accuracy of determination of a chance by sampling are greatly exaggerated.

In an examination of 60 questions, where marks are usually earned in multiples of about a third of a question, if a candidate knows a third of the whole syllabus, his expectation of units is 60 units; but the standard error for random sampling is

$$(180.\tfrac{1}{3}.\tfrac{2}{3})^{\frac{1}{2}} \doteqdot 6.4.$$

This is of the same order as the range covered by the second class; a candidate, if his actual ability is medium second class, would have appreciable chances of a low first or a high third. On the whole most people that get firsts are better than those that get seconds, and seconds than thirds, but the idea (held by many examiners) that an examination decides correctly for all of them is ridiculous.

3·6. *The 2×2 table.* This is possibly the commonest of all applications of statistical methods. Suppose we are considering two properties ϕ, ψ and want to know whether there is any association between them. We may sample a large population at random and count the numbers that have the four combinations of properties $\phi.\psi$, $\phi.\sim\psi$, $\sim\phi.\psi$, $\sim\phi.\sim\psi$. Alternatively we may have two sets consisting entirely of ϕ and $\sim\phi$ respectively. Take independent samples and count. In either case we get a 2×2 table. Let the numbers be

$$\begin{pmatrix} x & y \\ x' & y' \end{pmatrix}.$$

Let q be the hypothesis that there is no association between ϕ and ψ, that is, that the chances are in proportion. Then

$$K = \frac{P(q \mid \theta H)}{P(q' \mid \theta H)} \bigg/ \frac{P(q \mid H)}{P(q' \mid H)} \doteq \left\{ \frac{N^3(x+y)}{2\pi(x+x')(y+y')(x'+y')} \right\}^{\frac{1}{2}} \exp(-\tfrac{1}{2}\chi^2),$$

$$(1)$$

where $N = x+y+x'+y'$, $x+y$ is the least of the four row and column totals, and

$$\chi^2 = \frac{N(xy'-x'y)^2}{(x+y)(x+x')(y+y')(x'+y')}.$$

$$(2)$$

To test whether there is any association all four entries in the table are necessary. Most newspaper data and all advertisements give not more than one.

The definition of χ^2 is easily seen to agree with the earlier one. The actual values of the chances of ϕ and ψ do not say whether ϕ and ψ are associated, and if the counts are adjusted to make the row and column expectations the same but the entries in proportion, the first entry in the table would be $(x+x')(x+y)/N$, and

$$x - \frac{(x+x')(x+y)}{N} = \frac{xy'-x'y}{N}.$$

$$(3)$$

Other $O-C$ values differ only in sign; so that

$$\chi^2 = \left(\frac{xy'-x'y}{N} \right)^2 \left(\frac{N}{(x+y)(x+x')} + \frac{N}{(x+y)(y+y')} \right.$$

$$\left. + \frac{N}{(x'+x)(x'+y')} + \frac{N}{(y'+y)(y'+x')} \right), \quad (4)$$

which is easily seen to agree with (2).

The tests do not give the posterior probability of q completely, since it depends on the prior probability, to which we have not assigned a definite value. It would be desirable to be able to assign such a value, but so far as the general principles are concerned this is of secondary importance. The only difference between q and q' is that one parameter has a definite value on q and is left adjustable on q'; hence in any reasonable ordering of possible laws these laws will not be far apart in the series and will have prior probabilities of the same order of magnitude (differing, say, by not more than a factor 2). To make K large or very small a large number of observations is needed in any case; hence if K is large $P(q \mid \theta H)$ will be near 1, and if K is small $P(q \mid \theta H)$ will be small, with any reasonable assessment of the prior probabilities.

In Bayes's paper a uniform distribution of the prior probability of a chance was adopted. He explicitly stated that this was to be done only when there was no previous information about the value. His rule was adopted by Laplace. The amazing thing is that the rule was thenceforth treated as an article of religious faith. Supporters of Bayes and Laplace tried to use it in all circumstances whatever; opponents pointed out cases where it led to absurd results and rejected the theory entirely. Nobody tried to modify it until Pearson, who saw the problem but did not get far with the solution; and the first specific modification was made nearly 170 years after Bayes's paper!

3·7. At this point it is worth while to say something about the argument for induction, that it has always worked in the past and may therefore be expected to succeed in the future. The simple objection to this is that it is itself an inductive argument, and to use it to justify induction is a vicious circle. But this is not so; because, being about inductions, it would presumably be of a higher logical type than the inductions that it speaks about. It is impossible to prove deductively, of course, and would need a new postulate, but that is not a new problem. Another objection is that inductions have often failed in the past; and this seems to be the reason for the belief of some philosophers that Laplacian induction gives too high a probability for a general law, whereas I think that scientific method depends essentially on the opposite belief.

The fundamental objection to this solution is that, in its usual form, it postulates that general laws have been found to hold in the past. But the most that has ever been verified is that some general laws have had no exceptions hitherto (and it is very difficult to find such laws). It is not verified that any accepted general law will always hold in the future, and such a statement cannot be the basis of an induction even of higher type.

I think, however, that the argument can be restated so as to be informative. We have seen that on the Laplacian theory, if a sample of m members all have property ϕ, then there is probability $\frac{1}{2}$ that the next $m+1$ members will also have the property—and probability $\frac{1}{2}$ that they will not all have it.

Now suppose that we have many classes specified by different properties ϕ_r, and that we sample them according to further properties ψ_r. Suppose that the numbers of the samples that are all found to have ψ_r are m_r. Then in each case there is a probability $\frac{1}{2}$ of finding a not-ψ_r in the next m_r+1. Now these probabilities, referring to different defining properties for the parent classes and to many different ψ_r, are as nearly independent as we can reasonably hope. Thus Bernoulli's theorem should apply and we could infer that in approximately half the cases the generalization should break down within the next m_r+1 trials. This appears to be glaringly false. Much use has been made of the failure of the generalization 'all swans are white'; but its creator, presumably, made it from not more than a few hundred instances and the instances verified before the Australian black swan was found were many millions. Whether generalizations have continued to be verified or not, the ordinary state of affairs is that when a generalization has been found false the number of instances examined has been many times the number that it was originally based on. Then this can be made the basis of an induction of the second type, and leads to results similar to those given by the significance test for whether a chance is 0 or 1. We cannot predict how many ϕs are likely to occur if m have been observed with no not-ϕs before a not-ϕ is found; perhaps m^2 would be fairly representative. If one is found we can consider a more complicated hypothesis. But so long as no definite suggestion of one is made we cannot consider it.

This does, however, lead to the further question: is there a hierarchy of inductive types? What I am concerned with here is to state rules for the lowest type. But, for instance, the inverse square law of gravitation was first verified for the Earth and Mars; this created a presumption that it would also hold for Venus and Jupiter; and this being verified it was extended to the other planets. In application to each new planet the previous verifications created only a presumption, which had to be tested by observations of that planet. But now it is applied without question to any new asteroid or satellite. This appears to be an induction to analogous cases; some new principle, at any rate, is involved in the passage from a rule concerning two planets to one for a third, where the new observations do not form part of a continuation of the first two series. Further, Newton is said to have guessed the inverse square law for gravitation from its applicability to the intensity of light, and it was afterwards found to hold in electrostatics. Here there were suggestions that a law, verified in some fields, would hold in different fields altogether. I make no proposals here towards a solution of this problem, but merely point out that the problem exists.†

† I have discussed the problem in rather more detail in *Brit. J. Phil. Sci.* **5**, 1955, 275–89.

CHAPTER IV

ERRORS

A snapper-up of unconsidered trifles.
SHAKESPEARE, *A Winter's Tale*

4·0. The majority of the laws of physics and chemistry are of the form

$$y = f(x_1, x_2, \ldots, x_n)$$

where $y, x_1, \ldots, x_n$ are quantities determined by measurement and f is a given mathematical function. Such a law enables us to calculate y when the x's are given. These laws are established by repeated verification. It is found in numerous instances that the observed value of y agrees closely with that calculated from the law, and on the strength of the verification the law is accepted as general. The situation is similar to the problem of establishing a general law by sampling, which we considered in the last chapter, but there are several new features, which are usually obscured by confusion of language. Mach, for instance, is often quoted as having said that the choice of a scientific law is based on 'economy of thought', or 'economy of description', of observations. 'Description' is ambiguous, as we pointed out on p. 2. But the point is so important that further illustration is needed. We consider an experiment that is done in first year physics classes.

A solid of revolution can roll down an inclined plane, and its displacement is observed every fifth second after it starts from rest. Denote the time by t and the displacement by x. Suppose that the observations are as follows:

t (sec.)	0	5	10	15	20	25	30
x (cm.)	0	5	20	45	80	125	180

Then we can say that at all the instants of observation the displacement is connected with the time by the relation

$$5x = t^2. \tag{1}$$

On the face of it this statement is a pure description of observed facts. But the facts would be fitted equally well if the displacement was really connected with the time by the formula

$$5x = t^2 + t(t-5)(t-10)(t-15)(t-20)(t-25)(t-30)f(t), \quad (2)$$

where $f(t)$ might be any function whatever that is finite at

$$t = 0, 5, 10, ..., 30 \text{ sec.}$$

The law $5x = t^2$ is not the only description that fits the data; it is only one of an infinite number of descriptions that would fit the data equally well. Its special quality that distinguishes it from the other possible descriptions is its simplicity, but the importance of simplicity is not only a matter of economy of description of the existing observations. The law can also be used to predict the position of the body for other times of travel, and for any other time $f(t)$ could be chosen to give any position whatever. A physicist expects $f(t) = 0$ to give the correct value of x at other times than those so far observed, and would consider any other choice to be ridiculous.

Let us put the matter in another way. Some physicists would say that (1) is adopted because it is observed to be true. But what is observed is that for $t = 0, 5, 10, ..., 30$ sec., $5x = t^2$. This is merely a concise way of rewriting the observations, a shorthand description. What is asserted is that for *all* values of t, $5x = t^2$. This is an inference from a finite number of actual observations to an infinite number of possible observations. To express both by saying $5x = t^2$ is to use the same words to mean two different things.

This argument actually understates the position. In an actual experiment the observations would not fit the simple law exactly. We might get a set of values like the following:

t (sec.)	0	5	10	15	20	25	30
x (cm.)	0	5	19	44	81	124	178

These do not fit the simple law exactly, nor any other simple square law. But we could find a polynomial of seven terms

$$x = a_0 + a_1 t + a_2 t^2 + a_3 t^3 + a_4 t^4 + a_5 t^5 + a_6 t^6$$

that would fit the observations exactly. Nevertheless the physicist would still use the square law. His expressed reason would be interesting. It would be that *any* set of seven values whatever can be represented by an expression with seven adjustable constants. Consequently the expression so obtained could tell us nothing about the reliability of the determination. The very fact that the representation is of such generality that it can always be made to fit the data is regarded as an argument against it, not for it. With regard to the original square law, we would say that the observed values never differ from the calculated ones by more than 1 cm., except the last, and the difference could be accounted for by an error in timing of 0·17 sec., while the observations were made only to 0·2 sec. In fact he would say that the differences never exceed the admissible errors of observation, and that the agreement of the observations with the simple law is perfectly satisfactory.

We shall return to the physicist's stated reasons for his decision later. What we notice at once is that his predilection for the simple law is so strong that he will retain it when it does not satisfy the observations exactly, in spite of the existence of more complex laws that do satisfy them exactly. He would apply the law to predict the value of x for $t = 60$ sec. and would expect the result to be right within a few centimetres, provided the plane was long enough to permit the displacement required. He would, on the other hand, expect the polynomial of seven terms to give a seriously wrong result when extrapolated to such an extent.

The actual behaviour of physicists in always choosing the simplest law that fits the observations therefore corresponds exactly to what would be expected if they regarded the probability of making correct inferences as the chief determining factor in selecting a definite law out of an infinite number that would satisfy the observations equally well or better, and if they considered the simplest law as having the greatest prior probability. It is not explained by the reasons usually stated, which neglect the problem of inference completely.

The use of the expression 'exact science' and the scarcity of statements in popular works on physics about imperfection of agreement between physical laws and observation led most philosophers into thinking that exact agreement is attained. Where they mentioned errors of observation at all they dismissed them as a minor complication; and the uncertainty principle in the new quantum theories attracted an attention from philosophers out of all proportion to its novelty as a contribution to philosophy. Exact agreement between physical laws and observation was never attained; Plato knew this, and Laplace had a practical technique for dealing with it. The uncertainties treated in the quantum theory are far smaller than any of the discrepancies between previous theories and observational results; but these had attracted little attention from philosophers, some of whom even select Laplace as exemplifying extreme determinism.

The crucial point is that the observed values of x vary with t; the greater part of the variation is accounted for by the law, leaving an unexplained balance. The ground for accepting and generalizing the law is therefore a quantitative one: *how much* of the observed variation needs to be explained by the law before we can accept the law? The differences are called *errors* (in no derogatory sense) or *residuals*, and before we can proceed any further we must consider their properties.

4·1. *Measurement with a scale.* This is the simplest type of measurement. Suppose that an object whose length we want to know is placed with one end against the mark o on a scale. We read the position of the other end to the nearest multiple of the scale interval, which is the nth. Then the length lies between $n-\frac{1}{2}$ and $n+\frac{1}{2}$ scale intervals, and we can say no more. However often we repeat the experiment we shall always get the same result. Now suppose that the true length is x and that the prior probability that x lies in a particular interval dx is $f(x)\,dx$. The probability of reading n, given x, is 1 if $n-\frac{1}{2} < x < n+\frac{1}{2}$, otherwise zero, and this is true for any number of observations. Hence

$$P(dx \mid nH) = \frac{f(x)\,dx}{\displaystyle\int_{n+\frac{1}{2}}^{n+\frac{1}{2}} f(x)\,dx} \qquad n-\tfrac{1}{2} < x < n+\tfrac{1}{2}, \qquad (1)$$

$$P(dx \mid nH) = 0 \qquad\qquad x < n-\tfrac{1}{2}, x > n+\tfrac{1}{2}. \qquad (2)$$

If $f(x)$ does not vary much in the range $n-\frac{1}{2} < x < n+\frac{1}{2}$, however much it may vary outside that interval, the posterior probability of x is nearly uniformly distributed from $x = n-\frac{1}{2}$ to $x = n+\frac{1}{2}$. One observation tells us as much as any number would.

Suppose on the other hand that the true length is $n+x$, where n is a whole number and $0 \leqslant x \leqslant 1$. The object is now placed on the scale in an arbitrary position and the positions of both ends are read. The length is then found by subtraction. Suppose the first end is at $m+y$ with m an integer and $-\frac{1}{2} \leqslant y \leqslant \frac{1}{2}$. Then the second is at

$m+n+x+y$. The first end will be read as at m in any case. The other end will be read as at $m+n$ if $-\frac{1}{2}<x+y<\frac{1}{2}$, and $m+n+1$ if $\frac{1}{2}<x+y<\frac{3}{2}$. ($x+y$ cannot be $<-\frac{1}{2}$ or $>\frac{3}{2}$, so there are no other possibilities.) x is fixed, but in the conditions given the probability that y is between y_1 and y_2 is y_2-y_1 for $-\frac{1}{2}<y_1<y_2<\frac{1}{2}$. Then the length as measured will be n if $-\frac{1}{2}-x<y<\frac{1}{2}-x$; but since $x \geqslant 0$ the lowest member of this inequality can be replaced by $-\frac{1}{2}$, and the length of the permitted range for y is $1-x$. Thus the probability of reading n as the length is $1-x$. The measured length will be $n+1$ if $\frac{1}{2}-x<y<\frac{1}{2}$, and the length of the permitted range for y is x, so that the probability of reading $n+1$ is x.

Now take the prior probability distribution for $n+x$ to be $f(n+x)\,d(n+x)$. Suppose m readings to be made (the first end being placed afresh each time) and that l give n. Then the probability of getting n at one trial is $1-x$ and that of getting $n+1$ is x; and the chance of getting l of n and $m-l$ of $n+1$ is $^mC_l(1-x)^l x^{m-l}$ .for $0<x<1$, otherwise zero. Then the posterior probability that x lies in dx is

$$\frac{f(n+x)\,dx(1-x)^l x^{m-l}\,dx}{\int_0^1 f(n+x)\,dx(1-x)^l x^{m-l}\,dx} \doteqdot \frac{(m+1)!}{l!(m-l)!}(1-x)^l x^{m-l}\,dx \qquad (3)$$

if $f(x)$ does not vary greatly in a unit interval. (This condition will be satisfied, for instance, if we judge the object to be 10 cm. long within a centimetre or two and are measuring with a millimetre scale.)

If l and $m-l$ are large, (3) approximates to

$$\left\{ \frac{m^3}{2\pi l(m-l)} \right\}^{\frac{1}{2}} \exp\left\{ -\frac{m^3}{2l(m-l)}\left(x-\frac{m-l}{m} \right)^2 \right\} dx. \qquad (4)$$

The posterior probability of x is therefore nearly normally distributed about the mean of the observed values, which is $(m-l)/m$, with a standard error s_x given by

$$s_x^2 = \frac{l(m-l)}{m^3}. \qquad (5)$$

s_x decreases like $m^{-\frac{1}{2}}$ when the number of observations is large.

Where one end was always placed in contact with a scale division, the maximum error (without regard to sign) was $\frac{1}{2}$. Here the data

5

never completely exclude any value up to 1, since the posterior probability never vanishes in any range of x in $0 < x < 1$. So far the first method would appear to be the better. But with the first method the posterior probability distribution remains the same, however many observations are made; with the second, with enough observations we can establish practical certainty that the true value of x lies within some interval whose length is of order $m^{-\frac{1}{2}}$.

These two examples are given to show how the theory of probability works in the simplest type of measurement. In actual measurement we are not satisfied to read to the nearest multiple of the scale interval; we usually try to estimate to the nearest tenth. If we always succeeded in this the same argument would apply, since we could regard a tenth of the scale interval as the step of the scale we are using. But in fact we usually find, in a long series of readings, that more than two values occur. This is partly owing to personal peculiarities. Some people have a preference for estimating to even multiples of 0·1, while others may read 0·5 in preference to 0·4 or 0·6. But that is not all; in fact in actual measures the whole of the variation is *never* explained but has to be estimated from the observations themselves.

4·2. *The normal law of error.* In actual observations there are usually several sources of error besides the simple errors of reading that arise from the step of the measuring scale. In the problem of the body rolling down an inclined plane it was impossible to measure the time absolutely accurately, since the watch could not be read to less than 0·2 sec.; there was a possibility of inaccuracy in measuring the position of the moving object; the watch and the position of the body were not observed at exactly the same instant, since some time would elapse in looking from one to the other,† and possibly the slope of the plane was not exactly uniform. In astronomical observations disturbances come from variations of the temperature, vibration of star images (twinkling), unevenness of the micrometer screw, errors in setting the instrument, and a host of other

---

† With a stopwatch, which would normally be used, the measures would be made in a different way. Separate runs of the object would be taken to different distances; the watch would be started at the same time as the object and stopped when the object reached a given distance. This reduces some of the uncertainties, but it remains true that some time elapses between the object reaching a given mark and the pressing of the button that stops the watch.

phenomena. A good observer can make these very small by attending specially to the larger ones, but cannot remove them altogether.

The distribution of errors, when the quantity to be measured is as steady as possible, approximately follows the normal law. An idea of the reason for this can be given by a model, though it must not be regarded as a proof. Suppose that there are n possible sources of disturbance of any one observation, each equally likely to give disturbances of $\pm\epsilon$, and that they are independent. Then in a given observation the chance that l of them are positive, and therefore $n-l$ negative, giving an error $(2l-n)\epsilon$, is

$$^{n}C_l(\tfrac{1}{2})^l(\tfrac{1}{2})^{n-l}, \tag{1}$$

which approximates for large n to

$$P(l \mid n\epsilon H) = \left(\frac{2}{\pi n}\right)^{\frac{1}{2}} \exp\left\{-\frac{2}{n}(l-\tfrac{1}{2}n)^2\right\}. \tag{2}$$

If the error is x, an interval δx, large compared with ϵ, but small compared with $n^{\frac{1}{2}}\epsilon$, will contain $\delta x/2\epsilon$ values of l, and the probability that x will lie in an interval δx is approximately

$$\left(\frac{1}{2\pi n\epsilon^2}\right)^{\frac{1}{2}} \exp\left\{-\frac{x^2}{2n\epsilon^2}\right\} \delta x, \tag{3}$$

which is of the normal form with standard error σ given by

$$\sigma^2 = n\epsilon^2. \tag{4}$$

Criticisms of this model are that most sources of error are not restricted to just two values, and that the magnitudes of these values need not be the same for all sources. The approximation can be justified under much wider conditions, and is then known as the central limit theorem. It is enough that the components should all follow the same probability law and that this should have a second moment; and since the sum of two errors each following the normal law itself follows the normal law, the restriction that all the components must follow the same law can be relaxed. But it remains true that the conditions for the central limit theorem to hold are seldom *known* to be true in actual applications. Consequently the normal law as applied to actual observations can be justified, in the last resort, only by comparison with the observations themselves.

Such comparisons have often been made. The conditions of observation are kept as steady as possible, and the numbers of errors are classified in intervals. The numbers in the intervals are compared with those predicted by the normal law. The fit (judged by χ^2) is usually quite good up to about twice the standard error, but beyond that actual observations usually show too many large errors. These are not numerous anyhow; it usually takes 500 observations or more to give enough outlying ones to estimate the departure, but it appears that actual errors of observation are usually adequately fitted by a law of the form

$$P(dx \mid \sigma H) \propto \left(1 + \frac{x^2}{2m\sigma^2}\right)^{-m} dx \qquad (5)$$

with m from 2·5 to 8. $m = 4$ is typical. However, the difference is comparatively small and no allowance for it will be made here. We shall proceed as if the normal law held.

Suppose that we are trying to measure a quantity λ and that σ is known. The prior probability of λ is taken to vary smoothly over a range large compared with σ. Then

$$P(d\lambda \mid \sigma H) = f(\lambda)\, d\lambda. \qquad (6)$$

The probability, given λ and σ, that an observed value x will lie in a particular range dx is (since the error is $x - \lambda$)

$$P(dx \mid \lambda\sigma H) = \frac{1}{\sqrt{(2\pi)}\,\sigma} \exp\left\{-\frac{(x-\lambda)^2}{2\sigma^2}\right\} dx. \qquad (7)$$

Suppose then that, given λ and σ, the probabilities of values of x in different observations are independent. If the observed values are $x_1, x_2, \ldots, x_n$, the likelihood is got by multiplication, given the hypothesis of independence. Then

$$P(dx_1, dx_2, \ldots, dx_n \mid \lambda\sigma H)$$
$$= \frac{1}{(2\pi)^{\frac{1}{2}n}\,\sigma^n} \exp\left\{-\frac{1}{2\sigma^2}\Sigma(x_r - \lambda)^2\right\} dx_1 \ldots dx_n. \qquad (8)$$

If
$$n\bar{x} = \Sigma x_r, \qquad (9)$$

$$(n-1)s^2 = \Sigma(x_r - \bar{x})^2, \qquad (10)$$

we have, since $\Sigma(x_r - \bar{x}) = 0$,

$$\Sigma(x_r - \lambda)^2 = \Sigma\{(x_r - \bar{x}) + (\bar{x} - \lambda)\}^2 = n(\bar{x} - \lambda)^2 + (n-1)s^2. \qquad (11)$$

The intervals dx_r within which the observed values are recorded to lie are fixed by the conditions of observation (usually by the scale interval) and do not involve λ and σ; then by the principle of inverse probability

$$P(d\lambda \mid \sigma x_1, x_2, \ldots, x_n H)$$

$$\propto \sigma^{-n} f(\lambda) \exp\left[-\frac{1}{2\sigma^2} \{(n-1)s^2 + n(\bar{x}-\lambda)^2\} \right] d\lambda$$

$$\propto f(\lambda) \exp\left\{ -\frac{n}{2\sigma^2}(\bar{x}-\lambda)^2 \right\} d\lambda. \tag{12}$$

Hence if $f(\lambda)$ does not vary greatly in a range of λ of length σ, the posterior probability of λ is nearly normally distributed about $\bar{x}$ with standard error $\sigma/\sqrt{n}$.

Here the factor depending on s cancels and the estimates of λ and its uncertainty are independent of the scatter of the observations. This is a peculiarity of the normal law, and even for that it holds only if the standard error is given in advance.

If the standard error is not known in advance we must find a way of expressing its prior probability distribution. With no previous knowledge it could be arbitrarily large or arbitrarily small, but it must be positive. A clue can be obtained as follows. The normal law is now usually expressed in terms of σ, but in the nineteenth century it was usually written in terms of the precision constant h, thus

$$P(dx \mid \lambda \sigma H) = \frac{h}{\sqrt{\pi}} \exp\{-h^2(x-\lambda)^2\} dx \tag{13}$$

with $2h^2\sigma^2 = 1$. There is no strong reason for preferring h to σ, or vice versa, as a way of expressing the law, and for any interval σ_1 to σ_2 there is a corresponding interval h_1 to h_2 within which h must lie. Now in accordance with the principle that we want to take our rules in the most general form possible we should try to take probability distributions of the same form for σ and h, since there is no intrinsic restriction on either except that it is positive. This would say that there is a function g such that

$$P(h_1 < h < h_1 + \delta h) = g(h)\,\delta h \tag{14}$$

$$= P(\sigma_1 > \sigma > \sigma_1 - \delta\sigma) \propto g(\sigma)\,\delta\sigma. \tag{15}$$

But

$$g(h)\,\delta h \propto g(\sigma)\,\delta\sigma \tag{16}$$

is satisfied by

$$g(h) \propto 1/h, \quad g(\sigma) \propto 1/\sigma, \tag{17}$$

and by no other continuous function. Then the required condition of invariance is satisfied if we take

$$P(d\sigma \mid H) \propto d\sigma/\sigma. \tag{18}$$

The same argument applies in any case where it is equally natural to express a law in terms of some parameter or of a power of it. If the parameters are α and β, connected by

$$\beta = \alpha^m, \tag{19}$$

then
$$\frac{d\beta}{\beta} = \frac{m\,d\alpha}{\alpha}, \tag{20}$$

and if we take $P(d\alpha \mid H) \propto d\alpha/\alpha$, $P(d\beta \mid H) \propto d\beta/\beta$, we have equivalent forms. The constant factors can be determined if α is known to lie within stated bounds, and in any case cancel when posterior probabilities are calculated. Such a case is that of a density, where it is equally natural to express the property in terms of ρ, the mass per unit volume, or of v, the volume per unit mass. The rules $d\rho/\rho$ and dv/v are equivalent.

If we adhered to a uniform probability distribution for all unknowns we should be led directly to a contradiction, since dh is not proportional to $d\sigma$.

There remains the point that if

$$P(d\sigma \mid H) = A\,d\sigma/\sigma, \tag{21}$$

with any non-zero A,

$$P(0 < \sigma < \infty \mid H) = \infty, \tag{22}$$

thus contradicting the rule that probability 1 is to be attached to certainty. I have previously argued that this rule is a convention, and that we can consistently denote certainty by ∞ on specific data if it is convenient. The difficulty here is that if this is done for one certain proposition, on given data, it should be done for all, and then all ratios of probabilities become indeterminate; saying that we do not know a standard error would make us forget all that we knew about anything else.

If σ is known to lie between two values σ_1, σ_2, the rule (21) is applicable with

$$1/A = \log(\sigma_2/\sigma_1).$$

Suppose that σ_2/σ_1 is large, and that λ and σ give no information about each other. Then the combination of (6) and (18) gives

$$P(d\lambda\,d\sigma\,|\,H)\propto f(\lambda)\,d\lambda\frac{d\sigma}{\sigma}. \tag{23}$$

Combining this with (8) we have (θ denoting the observed values as a whole)

$$P(d\lambda\,d\sigma\,|\,\theta H)\propto f(\lambda)\,d\lambda\frac{d\sigma}{\sigma}\,\sigma^{-n}\exp\left[-\frac{1}{2\sigma^2}\{(n-1)s^2+n(\lambda-\bar{x})^2\}\right]. \tag{24}$$

The coefficient of $d\lambda\,d\sigma$ is large only for σ near s and λ near $\bar{x}$, and we can still neglect the variation of $f(\lambda)$ in this region. To get the posterior probability distributions of λ and σ we must integrate with regard to σ and λ respectively; then

$$P(d\sigma\,|\,\theta H)\propto\sigma^{-n}\exp\left\{-\frac{(n-1)s^2}{2\sigma^2}\right\}d\sigma \quad \sigma_1<\sigma<\sigma_2, \tag{25}$$

$$P(d\lambda\,|\,\theta H)\propto d\lambda\int_{\sigma_1}^{\sigma_2}\exp\left[-\frac{1}{2\sigma^2}\{(n-1)s^2+n(\lambda-\bar{x})^2\}\right]\frac{d\sigma}{\sigma^{n+1}}$$
$$\propto\{(n-1)s^2+n(\lambda-\bar{x})^2\}^{-\frac{1}{2}n}\,d\lambda \tag{26}$$

approximately, provided s is much greater than σ_1 and much smaller than σ_2.

If $n=1$, $(n-1)s^2=0$, and (25) reduces to $d\sigma/\sigma$, which was where we started. It states the sensible result that one observation can tell us nothing about its own uncertainty. (26) gives $d\lambda/|\lambda-\bar{x}|$, indicating that $\bar{x}$ is the most probable value but giving no indication of its accuracy. If $n=2$, the integrals of both expressions converge, and thus definite results are given from the smallest number of observations that we should expect to give any.

If n is large, (26) is approximately proportional to

$$\exp\left\{-\frac{n(\lambda-\bar{x})^2}{2s^2}\right\}d\lambda,$$

so that the posterior probability distribution of λ is nearly normally distributed about $\bar{x}$ with standard error $s/\sqrt{n}$. This rule, however, is badly wrong for small values of n.

The rule (26) was first given by 'Student'* by a different

---

* 'Student' was the pen-name of W. L. Gosset; his real name was announced publicly only after his death. Like Lewis Carroll, he is much better known under his pen-name.

approach, but is now usually stated in a form due to Sir R. A. Fisher. We call s the standard deviation of the observations, $s/\sqrt{n}$ the standard error of the estimate $\bar{x}$, and put

$$t = \frac{\lambda - \bar{x}}{s/\sqrt{n}}. \tag{27}$$

Then the rule becomes

$$P(dt \mid \theta H) \propto \left(1 + \frac{t^2}{n-1}\right)^{-\frac{1}{2}n} dt \propto \left(1 + \frac{t^2}{\nu}\right)^{-\frac{1}{2}(\nu+1)} dt, \tag{28}$$

where $\nu = n-1$. For n large this is nearly $\exp\left(-\frac{1}{2}t^2\right) dt$, and there would be a probability 0·68 that $-1 < t < 1$. But for $n = 2$ the corresponding probability is precisely $\frac{1}{2}$. This is equivalent to saying that, given two observations with no external information about their accuracy, the true value is as likely as not to lie between them. The difference is even greater for large values of t. For there to be a probability 0·01 that a given value of $|t|$ should be exceeded, $t = 2·58$ for n large; but for $n = 2$, $t = 63·66$. Consequently, if we are interested in the risk of large errors, it is most important that the number of observations should be given.

The extension to the estimation of m parameters (besides the standard error if not already given) is the method of least squares; it involves no new principle and I shall not describe it here. The constant ν is then $n - m$.

It is important to notice that so far as the normal law has anything to say about observed values, both the true value λ and the standard error σ are needed. The standard error is just as essential a part of the law as the true value.† To give no value of λ at all tells us nothing; to give λ with no indication of how much x is likely to depart from it tells us hardly any more. It is only when we have an additional parameter to specify the scale of the errors that we have anything useful to say.

The results for the estimation of λ and σ in the normal law are thus so far entirely sensible. We can now return to the question of the apparent divergence of the prior probability rule for σ when there are no previous bounds to its value. For $n \geqslant 2$ the integrals of (25) and (26) to 0 and ∞ converge, and it is natural to take the limiting values of the posterior probabilities as $\sigma_1 \rightarrow 0$, $\sigma_2 \rightarrow \infty$ as the correct

† I believe Fisher is responsible for this remark, but we have not succeeded in tracing it.

solutions when σ is unrestricted. This is consistent with the rule that certainty is denoted by probability 1. Thus the difficulty is avoided by the usual mathematical device for defining an improper integral.

The available information about λ and σ is completely prescribed when $\bar{x}$, s and n are known; but the prescription is in the form of the complete probability distribution. There is no question of picking out a particular pair of values as the most probable and saying that those are λ and σ. Some remarks of statisticians seem to indicate that they adopt the latter procedure, which has been seriously criticized by K. R. Popper.†

4·3. The rules for estimation of a parameter can be extended to any law of error, and except in a few special cases they take a standard form when the number of observations is large. We state this only for one parameter. If a law is

$$P(x \mid \alpha H) = \phi(x, \alpha) \tag{1}$$

and the prior probability for α is given by

$$P(d\alpha \mid H) = f(\alpha)\, d\alpha \tag{2}$$

then given n observations we have

$$P(d\alpha \mid x_1, x_2 \ldots x_n H) \propto f(\alpha)\, d\alpha \prod_{r=1}^{n} \phi(x_r, \alpha). \tag{3}$$

In general the product has a maximum with regard to α for a particular value a; and the maximum of the posterior probability density for α is given by

$$\frac{\partial}{\partial \alpha} \log f(\alpha) + \Sigma \frac{1}{\phi_r} \frac{\partial \phi_r}{\partial \alpha} = 0 \tag{4}$$

which is equivalent, for small $\alpha - a$, to

$$\frac{\partial}{\partial \alpha} \log f(\alpha) + \left(\Sigma \frac{\partial^2}{\partial \alpha^2} \log \phi_r \right)(\alpha - a) = 0. \tag{5}$$

When n is large, $\Sigma \dfrac{\partial^2}{\partial \alpha^2} \log \phi_r$ is large of the order of n; but $\log f(\alpha)$ is independent of n. Hence the value of $\alpha - a$ that satisfies (5) is small of the order of $1/n$. Further,

$$\frac{\partial^2}{\partial \alpha^2} (\Sigma \log \phi_r + \log f(\alpha)) \tag{6}$$

† *Brit. J. Philos. Sci.* **5**, 1954, 143–9.

is large of order n, and, since $\log f(\alpha)$ does not increase with n we have nearly

$$f(\alpha) \prod_{r=1}^{n} \phi_r \, d\alpha \propto \exp \tfrac{1}{2} \left[\Sigma \frac{\partial^2}{\partial \alpha^2} \log \phi_r \right] (\alpha - a)^2 \, d\alpha. \qquad (7)$$

The exponent is negative, since $\alpha = a$ gives a maximum, and is of order n. Thus for large n the estimate of α is nearly

$$\alpha = a \pm \left(-\Sigma \frac{\partial^2}{\partial \alpha^2} \log \phi_r \right)_{\alpha = a}^{-\frac{1}{2}} \qquad (8)$$

following approximately the normal law.

This argument was applied to sampling by Wrinch and me in 1919.[†] The main points are (1) that the uncertainty of the estimate of α is in any case of order $n^{-\frac{1}{2}}$; (2) that the difference between the values of α that make the posterior probability density and the likelihood maxima is of order n^{-1}, and therefore, for large n, is small compared with the uncertainty that is inevitable in any case; (3) that the posterior probability density is small except in a range of α whose length is of order $n^{\frac{1}{2}}$, and, provided that there is no violent irregularity of $f(\alpha)$ in this range, there is no important loss of accuracy if the posterior probability density is taken proportional to the likelihood. Subject to the last condition the precise form of the prior probability distribution is of little practical importance, though of course its theoretical importance remains.

4·4. Significance. A problem of significance is fundamentally one where the form of a law, on a null hypothesis, is specified; on the alternative hypothesis a parameter taken as zero on the null hypothesis is now taken to be adjustable. The significance test is obtained by sharing the prior probability between the two hypotheses, and on the alternative hypothesis giving the prior probability of the new parameter a continuous distribution. The problem of testing the correctness of a suggested value of a chance (3·5) is an instance on these lines, but the usual form can be seen as follows. Suppose that in the problem of estimating a true value, given the normal law and a known standard error, we had half the prior probability of

[†] *Phil. Mag.* **38**, 1919, 715–31.

α concentrated at $\alpha = 0$ and the rest smoothly distributed over a range of order σ, then we write

$$P(\alpha = 0 \,|\, H) = \tfrac{1}{2}; \quad P(d\alpha \,|\, H) = \tfrac{1}{2}f(\alpha)\,d\alpha, \quad (\alpha \neq 0), \qquad (1)$$

$$\int f(\alpha)\,d\alpha = 1. \qquad (2)$$

Then $\qquad P(\theta \,|\, \alpha H) \propto \dfrac{\sqrt{n}}{\sqrt{(2\pi)}\,\sigma} \exp\left[-\tfrac{1}{2}n\dfrac{(\alpha - \bar{x})^2}{\sigma^2}\right], \qquad (3)$

$$P(\alpha = 0 \,|\, \theta H) \propto \dfrac{\sqrt{n}}{\sqrt{(2\pi)}\,\sigma} \exp\left(-\tfrac{1}{2}n\dfrac{\bar{x}^2}{\sigma^2}\right), \qquad (4)$$

$$P(d\alpha \,|\, \theta H) \propto f(\alpha)\dfrac{\sqrt{n}}{\sqrt{(2\pi)}\,\sigma} \exp\left\{-\tfrac{1}{2}n\dfrac{(\alpha - \bar{x})^2}{\sigma^2}\right\}d\alpha, \quad \alpha \neq 0. \qquad (5)$$

If $f(\alpha)$ varies smoothly in a range about $\alpha = 0$, the latter leads to, nearly,
$$P(\alpha \neq 0 \,|\, \theta H) \propto f(\bar{x}), \qquad (6)$$

whence $\qquad \dfrac{P(\alpha = 0 \,|\, \theta H)}{P(\alpha \neq 0 \,|\, \theta H)} \doteq \dfrac{\sqrt{n}}{\sqrt{(2\pi)}\,\sigma f(\bar{x})} \exp\left(-\dfrac{n\bar{x}^2}{2\sigma^2}\right). \qquad (7)$

If $f(\bar{x}) \neq 0$, and $|\bar{x}| \leqslant \sigma/\sqrt{n}$, this ratio is large of order $\sqrt{n}$ for large n and there is strong support for the null hypothesis. If, however, $\bar{x}$ is large compared to $\sigma/\sqrt{n}$, the exponential factor swamps $\sqrt{n}$, and there is strong support for $\alpha \neq 0$.

Analysis of this type has been applied in detail to most of the standard problems where significance tests are needed, but the above outline indicates the general character of the results. The main point is that the null hypothesis is in general strongly supported if the maximum likelihood estimate of the new parameter is less than its standard error; but the introduction of the new parameter is strongly supported if the estimate is much more than the standard error. With ordinary numbers of observations (from 20 to 1000) the transition comes at about 3 times the standard error in most problems.

This analysis shows that the problem considered at the beginning of this chapter receives a satisfactory answer from the theory of probability. If we take as null hypothesis the proposition that the body has a constant acceleration, we can compare it with the alternative hypothesis that x is of the form $\tfrac{1}{2}ft^2 + \alpha t^3$, where α is an adjustable parameter. Then the data lead to a maximum likelihood estimate of α, with an uncertainty estimated from the residuals. If this estimate is large enough in comparison with the uncertainty it establishes a high probability that the correcting term is needed. If it is smaller than the uncertainty (without needing to be zero) the data support the hypothesis that no such term is needed.

It must be remarked that physicists hardly ever give a statement of uncertainty derived from the observations in the way required. If they give any it is usually little more than a guess, and quite useless for a significance test. A proper estimate of uncertainty can be recovered from the original readings, if these are published, but more often they are not. A physicist will cheerfully spend months on an experiment and grudge the few hours needed to present his results in the form that would make them of full use to other people.

In astronomy, estimates of uncertainty are usually given, but usually without mention of the number of observations. This information, when the number of observations is large, is enough in an estimation problem, but if there is any question of significance of the difference between two estimates the number of degrees of freedom becomes essential.

4·5. The use of the simplified form of the product rule in the above analysis supposes that, given the parameters, the probability of a given error in one observation is completely independent of those of previous observations. This is not necessarily true. It is possible for a large component of the error to keep the same sign over several successive observations. If there are m groups of p observations each, and the standard error of one observation is σ, but if the actual error is the same for all observations of a given group, the mp observations give no more information than m independent observations would. Less extreme cases can be dealt with in practice. If the observations were wholly independent, the standard error of a group of p would be $\sigma/\sqrt{p}$. Tendency of errors to recur would increase this; but if the actual group means are determined it is possible to apply a significance test to see whether the scatter is consistent with the hypothesis that it is derived from a standard error $\sigma/\sqrt{p}$ or something larger. If the latter hypothesis is supported, a standard error of the general mean can be found from the scatter of the group means. Errors that persist to some extent over several observations were called semi-systematic by Newcomb. Personal error of observation is a case in point. Pearson showed that successive errors were sometimes correlated in such a way that the mean of a series of 500 observations might be no more representative than the mean of 50 to 300 observations of the same scatter.†

Errors that could be precisely calculated for each observation, given the values of certain parameters, are called systematic errors. In the dip-circle method of determining the inclination of the Earth's magnetic field several such errors are taken into account, one being the angle between the geometrical and magnetic axes of the needle. This produces equal errors in opposite senses when the needle is turned over, and is eliminated by taking the mean. The skill of an observer consists largely in detecting and eliminating such errors.

At present errors are treated very differently in different subjects. On the whole the highest standard is reached in biology; this is due mainly to the work of Pearson and Fisher. The actual error of observation here is usually small compared with the whole amount of variation. Thus in a test of a fertilizer the weights found for the crops on different plots, even with the same treatment, differ far more on account of real differences in the soil and pest attack than on account of error in weighing. The error is then mainly of the semi-systematic type. The plots receiving different treatments are arranged so that any such error will contribute at random to the plots receiving any particular treatment; thus the semi-systematic error is converted into a random error.

In astronomy the standard error of a single observation of position is usually of the order of $0 \cdot 1''$, and far less than the usual range of variation. Simple laws usually account for most of the variation, leaving a balance which is comparable with the error of observation. However, systematic variations are usually found when the best fit by a simple law has been found, and discoveries of far-reaching importance have arisen from the further analysis of these variations. The problem of significance is then of the first importance. Small oscillations of the Earth's axis, with amplitudes of the order of $0 \cdot 1''$, give valuable information about the structure of the Earth; small disturbances of the motions of satellites tell us much about

† *Phil. Trans.* A, **198**, 1902, 235–99; H. Jeffreys, *Phil. Trans.* A, **237**, 1938, 231–71.

the structures of their primaries and enable us to test theoretical suggestions. Such phenomena have to be shown statistically to be genuine. If the standard error of one observation is o·1″ and we want to measure a quantity of o·01″, it is necessary to reduce the standard error of the estimate to below o·005″. This will need 400 observations in the most favourable case; but if a systematic error is liable to contribute o·005″ the whole programme would be ruined.

In many branches of physics the position is much less satisfactory. The fitting of laws to the data is still often done graphically, and this introduces unknown personal errors, while estimates of uncertainty, if any are given, are usually based on 'judgment', that is, guesswork. This is very largely due to a misconception of the meaning of a standard error, based on the fact that when technique is improved estimates are sometimes found to have been wrong by more than their standard errors. (The larger class that have been right is less often mentioned.) In fact the probability that an estimate will be right within its estimated standard error is never more than $\frac{2}{3}$ (less if the number of observations is small); but the standard error expresses a scale for the probability of errors *on the hypotheses used in estimating it*. We are never sure that these hypotheses are right; they may be, but in any case comparison with other data is a valuable means for testing mistakes in hypotheses. But such comparison is impossible unless we know the standard errors. I have often found that alleged discrepancies were no more than could easily be attributed to random error, and that other discrepancies have been ignored, though they were many times the standard errors, because they were less than inflated uncertainties based on judgment. There is no harm in giving possible bounds for a systematic error, but so long as there is no way of finding its actual value it remains possible that somebody will succeed in evaluating it and correcting for it, and then the straightforward standard error will be right for the corrected value. The essential point is that the standard error is a minimum estimate of uncertainty; and, unless we know this minimum, comparison of estimates is futile.

4·6. In spite of the unsatisfactory features of most presentations of observational results, we usually have to do our best with the presentations available. What can reasonably be inferred from them? Suppose, for instance, that the observations concern the variation of y and x, and that the greatest and least values of x differ by a, those of y by b. Suppose further that the number of observations is large and that a linear form fits the values of y within $\pm c$, where c is positive and much less than $|b|$. Then the significance test for constancy of y with variation of x would show strongly that y does depend on x. Further, though the material might not be in a form suitable for testing forms that complicate the linear law, it is enough to show that any departure from the linear law is not enough to produce disturbances greater than c when x varies by a. If such a departure increases steadily with x, it could be detected ultimately by using a longer interval of x. Thus many of the main results could ultimately be found without using the detailed theory of probability. The use of this theory does, however, serve to increase the accuracy of the determinations in an estimation problem and to give decisive results in significance tests with a smaller number of observations. Further, the crude method would never detect systematic variations small compared with the standard error of one observation; and many important results depend on such variations.

The general principle that scientific method consists of a process of successive approximation is inevitable . The alternative would be that before we could say anything about a phenomenon we would have to know everything about it. Thus in treating a series of measures we may consider possible systematic variations one at a time, the hypotheses compared at each stage including the supposition that the outstanding variation is random. If we did not use the method of successive approximation we should be committed to explaining in detail every separate residual, and this is quite impracticable.

The loose use of probability in everyday life involves many further complications. One is that we habitually have to make decisions on data that are inadequately recorded. Another is that a practical decision must usually be made within a specified time; if this time is less than the time needed to calculate the probabilities concerned we cannot use the theory at all. But there are many cases of high or low probabilities where the full theory would give results in accord with common usage; jam does not often taste of onions; a train seldom starts before the time announced in the timetable; I can depend on walking a mile in twenty minutes; and so on. The difficulties for intermediate probabilities could arise equally seriously in applying deductive logic.

CHAPTER V

PHYSICAL MAGNITUDES

Multiplication is vexation;
Division's just as bad;
The Rule of Three perplexes me,
And Practice drives me mad.

NURSERY RHYME

5·0. So far we have been considering the possible ways of establishing scientific laws with high probabilities, and have shown how the theory of probability can achieve this. We shall proceed to the relations between some of the chief laws of physics and the observational evidence that supports them. Before doing so, however, we must comment on some features in the usual presentations in textbooks. These start by stating a set of laws, taken as fundamental, and proceed to develop a series of consequences; but the observational evidence is usually presented only incidentally. This may be inevitable in elementary teaching; there is no doubt that the theory of probability is rather difficult, and certainly more difficult than deduction. However, the usual presentations are unsatisfactory because they take no account of errors, and if the results were compared with observation directly, almost every law would be rejected at the next trial. The observed value hardly ever agrees exactly with prediction, and the laws are taught as if they were exact. This is as much of the so-called principle of causality as survives in applied mathematics, and it is essentially inapplicable unless it is supplemented by the notion of error; and when this is done the laws of physics, so far as they relate to observations, become statements of probability distributions. The quantities of physics arise fundamentally as parameters in these probability distributions.

We should mention at once what has been called the 'hypothetico-deductive method'. This consists in stating laws for consideration, comparing with observation, and rejecting them if they disagree too badly. Then modifications are suggested and the process is repeated. This process is, of course, an essential part of the method described so far. What is unsatisfactory about it is its incompleteness. Since errors of observation exist, any criterion for rejection must be quantitative; and there is no such criterion without at least a partial theory of probability. Further, what, if anything, is being said about a hypothesis when it is rejected? On the present theory what is said is that the hypothesis is so improbable, given the observations, that it contributes a negligible probability to any further inferences that might be drawn. But the statements of the hypothetico-deductive method mostly avoid speaking of the probability of a hypothesis and thus fail to deal with the problem of prediction.

On the present view the classical view of causality is inverted. Instead of saying that every event has a cause, we recognize that observations vary and regard scientific method as a procedure for analysing the variation. Our starting-point is to consider all variation as random; then successive significance tests warrant the treatment of more and more of it as predictable, and we explicitly regard the method as one of successive approximation.

The quantities that occur in physical laws appear in various ways, but there are three main types. Some are pure numbers. Some are found by direct measurement, such as length and duration of time. Others arise in stating relations between results of measurement and are usually ratios. The distinctions were treated at considerable length by N. R. Campbell in his *Physics: the Elements*. In the following account I have been greatly influenced by Campbell's writings, though at some points I disagree with him.

5·1. A fundamental notion of any science is number. In its most elementary form this means the number of members of a class, and depends on the notion of comparison of classes by pairing their members, as described in Appendix I. Similar classes are said to have the same number; this is a property of the classes, and is not shared by any class not cardinally similar to them. If two classes have no common member, and we take the class formed by taking all members of both, the number of this class is said to be the sum of the numbers of the original classes. If we form the class of all possible pairs of members of the two classes, the number of this class is called the product of the numbers of the original classes. If a class has no member its number is 0. If when a is a member of the class every member is identical with a, the number of the class is called 1. If a unit class is combined with a different unit class, the resulting class is said to have the number 2, and so on. In this way the finite whole numbers can be defined, and arithmetic can then be developed.

Number is an abstraction. When classes are similar we say that they have a common property and call this their number. The observed fact is the result of the comparison; the property is an abstract idea derived from it. This derivation by abstraction is a logical step, and is of extremely wide application. We experience a similar sensation from the sight of blood, a brick, a sunset, and a tomato; we abstract the common property, which we call redness, and which is not shared by the midday sky, a lemon, or grass. All qualifying adjectives depend for their meaning on such processes, of different complexities in different cases. In such an expression as 'ten men', 'ten' is not an adjective qualifying 'men'; this is seen at once if we try to attribute a meaning to 'a ten man'. 'Ten' here qualifies a class of men; 'ten men' means 'every man in a ten-class of men'. Sometimes, when objects are classified in terms of some method of comparison, the classes can be arranged in some definite order suggested by the method of comparison itself; thus we attach meanings not only to the statement that classes have the same number, but to the statement that one class has a greater number than another, and this makes it possible to arrange numbers in a definite order. This is a necessary requirement of a physical magnitude. It is not possessed by all abstractions. For instance, we can classify objects according to the colour sensation they give. But there is no direct reason suggested by our method of comparison according to colour to say what should be the order of arrangement of red, yellow and brown. For this reason colour is not a physical magnitude. In the case of the pitch of a note, we can say directly from sensation that one note is higher than another, and all audible notes can be arranged in a single series based on this comparison. Something more is needed, however, before we can *measure* pitch. The existence of an order is necessary to measurement, but other conditions must be satisfied before we can make a quantitative determination.

The further theory of numbers generalizes the idea of number in various ways. If we keep to positive integers we find difficulties like the following. In general, two numbers have a sum and a product.

If $ab = c$, we can think of b as the number that, multiplying a, will give c, and write $b = c/a$. Similarly if $a + d = c$ we can introduce the operation of subtraction and write $d = c - a$. But these ideas give a value to b only in cases where c is an exact multiple of a, and to d when c is $\geq a$. It took mathematicians a long time to see that meanings can be given to the operations of division and subtraction when these conditions are not satisfied; I shall not discuss the purely mathematical definitions here, but in the first place they were largely suggested by physical and economic needs. The important points for our purposes are that by introducing negative numbers we can make a system of numbers that permits us to subtract any number from any other; and by introducing rational fractions we can make division always possible without going outside the system.

However, even this was not enough for the needs of geometry. By purely geometrical methods the Greeks proved that the area of the square on the diagonal of a square was twice that of the original square. Supposing that both the side and the diagonal were rational multiples of a unit, it would follow that there are two whole numbers a, b satisfying the condition

$$a^2 = 2b^2.$$

If a and b have a common factor, we can divide out by its square, and thus we shall have the same relation satisfied by p, q, where p, q have no common factor, and

$$p^2 = 2q^2.$$

Since p^2 is divisible by 2, p must be divisible by 2; put $p = 2r$ and divide by 2. Then

$$q^2 = 2r^2$$

and q also is divisible by 2, contradicting the supposition that p and q have no common factor. Thus the rational numbers were not enough for geometry. This result placed the Greeks in a difficult position, which they dealt with by abandoning measurement in terms of a unit. The modern method depends on the opposite decision; we make a further generalization of number to admit $\sqrt{2}$. The Greeks had further difficulties; they tried to express $2^{\frac{1}{3}}$ and π (to translate into modern language) in an extended system that admitted square roots of all whole numbers, but failed. It is now known that such expressions are impossible. But it is also known that a further extension of the number system makes it possible to include both $\sqrt{2}$, $2^{\frac{1}{3}}$, and π. This is not the place to explain it in detail, but the essential idea is that of the Dedekind cut. If we have two sets of numbers a, b such that every a is less than every b, even if the as and bs are restricted to rational numbers, either (1) the as have a greatest member, or (2) the bs have a least member, or (3) the as have no greatest member and the bs no least member. If further the as and bs* are such that, for any positive ϵ, however small, there exist an a and a b such that $b - a < \epsilon$, the gap between the set of as and the set of bs is arbitrarily narrow and we can invent a new number to occupy it. This is what we call a real number. The real numbers have the following two properties. (1) The fundamental operations of addition, subtraction, multiplication and division can be applied to them, being first defined in terms of the operations for rationals. These operations always give a real number, except that we must not divide by 0. (2) Further attempts to extend the real number system by the method of classification applied to the rationals give only real numbers. That is, the real numbers are the most general set that admit the four fundamental operations and can be put in an order by the relation 'greater than'. Then $\sqrt{2}$, $2^{\frac{1}{3}}$ and π find suitable places among the real numbers.

There are various attitudes toward the logical status of the real numbers and their manipulation. In physics we are concerned chiefly with their use, and the question may be asked whether we need such generality as they provide. It is certain that we shall never need all the real numbers to express what we need to say; the need for them arises from the fact that if our theory was applicable only to some more restricted set we should have to examine in every operation whether the result is within that set. With the real numbers we know that it is.

5·2. *Additive magnitudes.* These are directly measurable quantities characterized by the possibility of a physical process of addition. 'Fundamental magnitudes' was the term used in Campbell's book, but he later recommended 'elementary'. Both terms are liable to misinterpretation; we are here considering the beginnings of a system of physics, whereas 'fundamental magnitudes' suggests to many such quantities as the quantum and the mass

---

* We do not indicate separate as, bs by suffixes, because use of a_n would suggest that the a's are an enumerable set; this is not essential to the argument.

of the electron, which appear at a much later stage. 'Elementary' suffers from the same drawback, but perhaps not so seriously.

We consider first one of the most important, namely length. When two objects can be placed so that they are in contact at both ends, we find by experiment that calipers or compasses adjusted so that they fit one object will also fit the other. Objects can therefore be classified together if they fit the calipers when the calipers are kept in the same adjustment. We abstract the common property, which we call the length of the objects. But the method of comparison by juxtaposition of the objects, either directly or by way of the calipers, suggests a way of arranging them in an order. If the calipers, having fitted one object, have to be extended to fit another, we say that the second has the greater length. The method of comparison not only gives a meaning to length, but arranges lengths in order, so that any length is greater than any that precedes it in the order and less than any that follows it. Thus length, so far, is on the same footing as the pitch of a note. The principles used up to this point are only those used in the theory of sampling. We have, however, an ordering relation among lengths as among numbers. But we can go further.

Consider the method of construction of a millimetre scale. A long screw is fixed so that it can turn in a bearing with a screw thread inside it. Whatever part of the screw is within the bearing, it fits. Every turn of the screw fits any turn of the bearing. In terms of our method of comparison, every turn of either therefore has the same length. In the manufacture of a scale, it is arranged that, whenever the screw advances through a complete turn, a device attached to it rules a transverse line on the scale. The object whose ends are two consecutive scale-divisions is therefore compared directly with the turn of the screw, which is known to have always the same length. Hence by the very definition of length every interval between consecutive divisions on the scale has the same length.

When we measure a length we place the ends of the object in contact with the scale or we apply calipers to the ends and apply the calipers to the scale; and we count the scale-intervals between the ends. The statement that the length of an object is 153 mm. means then that the object has the same length as the object formed by placing 153 scale-intervals end to end, all the intervals by construction having the same length as the turn of a standard screw. We see now the difference between a length and the pitch of a note.

When we put two objects together end to end along a scale we get a new object determined by the extreme ends. In terms of our method of measurement, merely by counting scale-intervals, the combined object has a length equal to the sum of those of the separate objects. But if we sound two notes together we do not get a single note of a new pitch. If the notes are sounded together we get a chord; if they are sounded in succession we get two distinct notes.

We can now specify in what conditions a property can be an additive magnitude. It must be possible to construct a scale such that every interval of the scale is the same in respect of that property, the test being comparison with some definite standard by the process that enables us to recognize differences in the property. The intervals must be consecutive, and the measure is the number of intervals overlapped. A rule of addition holds.

When the length of an object is an exact number of millimetres there is no essential difficulty about the addition rule. But suppose two objects have lengths 153 and 154 mm. and we read on a scale, reading only to the nearest centimetre. We shall read both lengths as 15 cm. The combined object, however, has length 307 mm., which we shall read as 31 cm.; and $31 \neq 15 + 15$. Similarly if we make a habit of reading to the nearest millimetre we may find that the addition rule is sometimes wrong by 1 mm. If we maintain the addition rule for measurement of lengths in all circumstances, we cannot retain universally the rule of measuring to the nearest scale division and considering the result as the length; and conversely, crude lengths measured in this way do not always satisfy the addition rule. What are we to do? Those who maintain that all physics should rest on actual measures and nothing else should apparently adopt the second alternative and abandon the addition rule for lengths. I do not know whether anybody does so, but most physicists maintain the addition rule and extend the meaning of length. This is now regarded as an intrinsic property of the object, to which actual measures are approximations. There is some direct reason for this. The failure of the addition rule when the measures are made to the nearest centimetre is largely explained if we read to the nearest millimetre; but the addition rule still sometimes gives discrepancies of a millimetre, most of which disappear if we estimate to tenths of a millimetre. At this point we are usually unable to read more closely, but we attribute this to an imperfection of our own vision. We may extend our accuracy further by means of a travelling microscope or an interferometer; but having got so far we still have not reached measures that always satisfy the addition rule. But what we can do is to regard the actual measure as the sum of two parts, one being the true length and the other the error; and by adopting the rule of measuring by difference and taking many measures we can estimate the former as closely as we like. Then the addition rule can be tested by a significance test. The true length can be regarded as an idealized extrapolation from actual measures; but in estimating it we treat it simply as a parameter in a certain probability law. Incidentally, since no such law is necessarily final, we can admit the possibility that the addition law might need modification at very great or very small lengths if occasion should arise.

5·3. *Scales and units.* Our result so far is of the form 'the length of an object, in comparison with a millimetre scale, is associated with a number $x \pm s_x$'. In fact what we usually say is 'the length of the object is $x \pm s_x$ millimetres'. This is intended to mean precisely the same, but is stated in what we shall call the language of quantity. A length could, in this language, be 1 mm., so '1 mm.' is a property of an object, namely its length. It seems queer that we should appear to multiply a property by a number, but this is precisely what is done in ordinary multiplication. A cardinal number n is a property of a class; if we take m classes of number n, with no common member, the number of the class that includes all their members is mn. Even in this case we need to take care that the classes have no common member; so it is not surprising that when we say that the length of an object is $x \pm s_x$ mm. we are not thinking of comparing it with a lot of detached millimetre objects, but with a scale formed of such objects end to end. So far there is no genuine objection to the language of quantity, though it can be regarded as simply an abbreviation.

We are not compelled to use a millimetre scale; we might for instance use one graduated in tenths of an inch. Such a scale can be compared directly with a millimetre scale, and we find that the length of 100 intervals on the tenth-inch scale is *the same* as the length of 254 intervals on the millimetre scale; it is the common property revealed by the method of comparison. Then, from the additive property of lengths we infer that if the length of an object, in comparison with a tenth-inch scale, is associated with the number x, it will be associated with $2 \cdot 54x$ when compared with a millimetre scale. This can of course also be verified directly. In the language of quantity this becomes simply 'x tenths of an inch $= 2 \cdot 54x$ millimetres'. The factor $2 \cdot 54$ provides a rule for converting a measure from one unit to the other.†

We have had two additive magnitudes so far, cardinal number and length. Angle, as measured by a protractor graduated in degrees, is another, for each degree-interval is compared with a standard length in the construction of the instrument. Angle can, however, also be defined in terms of ratios of lengths. Time, or rather the interval of time required for a given process, is another. It is measured by counting the swings of a pendulum or a balance wheel, which occur in a definite order, so that each has an immediate successor, and this order is recognized directly by sight or sound. When two processes, started at the same instant, also end at the same instant, we classify them together and abstract the common property, the interval of time taken. We measure this by counting the number of oscillations of a standard instrument, say a seconds pendulum, the balance wheel of a watch, or a tuning fork, that take place during either process. It is important that the interval is independent of the actual instant when the process starts, just as a length measured on a scale is independent of the position on the scale of the end first placed in position. By its essential structure, interval of time is therefore an additive magnitude. Time may also be measured in terms of the rotation of the Earth. The interval taken by the Earth to turn through a standard angle provides the scale, and any interval is measured as the multiple of this angle that the Earth turns through during the process. Angle being an

---

† Some unfortunate linguistic usages are prevalent. One, particularly associated with Professor Eddington, is that physical measures are simply numbers; which reminds one of early lessons in algebra when somebody wrote 'let $x =$ the cows'. Eddington's casualness about stating units did not prevent him from restoring them when they were needed for comparison with observation; and he could remark that the mass of the Sun is $1 \cdot 5$ kilometres without explaining just how this result was found.

A commoner expression is of the form 'The length of A, when measured in centimetres, is $17 \cdot 7$'. But $17 \cdot 7$ is a number; and whatever a length is it is not a number; and also, whatever a length is, it is the same whatever scale we use for comparison.

additive magnitude, interval of time measured in terms of it is another.†

Mass, as found on a balance, is another additive magnitude. The bodies we call our 'weights' are constructed so that they all counterbalance the same body on the other pan. We can recognize when a body more than counterbalances, or fails to counterbalance, a body on the other pan. We classify together bodies that counterbalance the same body, and abstract the property of mass. In this case there is a slight complication. We do not in practice weigh in terms of milligram weights alone. We use weights found by experiment to be equivalent, in regard to objects counterpoised by them, to various multiples of the unit. This process is analogous to measuring a length in terms of decimetres, centimetres and millimetres and using the known standards of comparison to convert the whole to centimetres.

5·4. *Derived magnitudes.* The majority of physical magnitudes are not measured directly. They occur as factors of proportionality in laws. Probably the only laws that do not involve such factors are those of simple constancy and those that express addition of measures of additive magnitudes in terms of the same scale. Nevertheless they may be connected with properties. For instance, liquids may be classified according as a given solid floats or sinks in them, and this relation is unaffected by the size and shape of the containing vessel, so long as the solid does not actually touch the sides. Using different solids we can divide liquids into two classes in terms of each. This method establishes an order among liquids and solids. It is found that if we have made a classification in terms of one solid and then try another, the second either floats in all the liquids that the first floats in, or sinks in all those the the first sinks in. There may be an intermediate group such that one solid floats and the other sinks. The liquids may therefore be arranged in an order such that each supports all solids supported by those before it in the series, but will not support some solids supported by liquids after it in the series. Then each liquid is said to have a greater density than those that precede it and a smaller one than those that follow it.

† Actually the Earth's rotation shows slight fluctuations and in the most accurate astronomical work the revolution about the Sun is considered to give a better standard for long intervals, and a crystal or atomic clock for short ones.

We have abstracted from the empirical relation the property of density. The process resembles in outline that of abstracting the notion of length from the results of juxtaposition. But the analogy breaks down at the next stage. We cannot construct a scale for measuring density by combining objects. In dealing with length we could put two millimetre intervals in succession and call the length of an object that fits the two together 2 mm. In dealing with time a process such that a seconds pendulum swings twice during it is said to occupy 2 sec. In each case two of the standard intervals together are greater, in terms of the method of comparison, than either separately. It is this fact that makes it possible to construct a scale. But in the case of density, when we put together two of the solids used for comparison, the combined solid does not determine a cut in the series of liquids outside those determined by the two solids separately; in general the cut it gives lies between the two former ones. There is no way of constructing a scale based on a single solid; and the measurement of density as an elementary magnitude breaks down.

We can, however, proceed in another way. We can weigh a portion of a liquid on a balance, and we can find its volume by means of a measuring glass. Both volume and mass are elementary magnitudes, and when the process is carried out several times on different portions of the same liquid it is found that they are proportional; that is, they are connected by a relation of the form

$$y = Ax,$$

with A an adjustable constant. At present we regard x and y as simply the numbers that appear in the measures, the scales being kept the same. Different liquids give different values of A, which thus expresses a property of the liquid; thus we can arrange liquids according to the values of A that they give. We also find that the order of increasing values of A is the same as the order specified by the results of flotation experiments. We then have a quantitatively specified value, the mass per unit volume, such that the greater mass per unit volume corresponds completely to greater or less density. In this way we associate a number with a density.

Density is an example of a derived magnitude. It can be ordered, but not directly measured in terms of a single scale. A series of experiments must be conducted, such that in each experiment two elementary magnitudes are measured; and the measures are then found to be connected by a simple relation, in this case proportionality. The constant factor is called the density. In the simple case of comparison of two scales of measurement we have already introduced a derived magnitude, by saying that the length of an object is 2·54 mm. for every tenth of an inch. Here we began by establishing a rule of proportionality valid for measures on the two scales, and found the number 2·54 as the actual one correct for the particular scales chosen. But its character is less evident than for density because the properties it enables us to compare are merely two different ways of specifying the same thing, namely the length of a given object. In the case of density the derived magnitude provides a way of connecting two quite distinct properties of a portion of liquid, namely its volume and its mass.

A derived magnitude associates a property with a number and one or more units. The number associated with a density is different according as the mass is measured in grams or in pounds, and the volume in cubic centimetres or cubic feet. When we say that a density is 1·34 grams per cubic centimetre, the expression '1·34 grams per cubic centimetre' must be taken as a whole; no item in it, neither '1·34' nor 'grams' nor 'cubic centimetre' can be changed without altering the meaning of the whole. For this reason it is misleading to speak, as is often done in writings on the theory of dimensions, of a 'mere change of units'. There is no such thing as a mere change of units. If we alter a unit without altering the number in the measure, we are speaking of a different physical system, and cannot assert anything about it without a physical law to guide us; while if we already know the physical law, a change of units tells us nothing that we cannot find out by keeping the same units and altering the numerical measure.†

In discussing length we began with length as a property of an object and led up to the idea of length as a quantity. Between the two a one-one correspondence exists. If two objects differ in the property, as tested by direct juxtaposition, they have different measures, and conversely. Similarly for density there is a one-one correspondence between the property tested by flotation and the mass per unit volume. This suggests that we may be able to treat

† For this reason the so-called method of dimensions is fallacious. It should be replaced by that of similarity, as Campbell has explained.

density also as a quantity. In the equation above, suppose that the numbers x and y were associated with a specimen by means of a cubic centimetre measure and a set of gram weights. Then in the language of quantity we say that the volume was x cubic centimetres and the mass y grams; but these are complete expressions of the quantities and may be denoted by v and m. Then the equation $y = Ax$ can be rewritten

$$\frac{m}{1 \text{ gm.}} = \frac{Av}{1 \text{ c.c.}}. \tag{1}$$

But m is proportional to v, and if we write $m = \rho v$ (1) is formally correct provided

$$\rho = A \frac{1 \text{ gm.}}{1 \text{ c.c.}}. \tag{2}$$

A is a number; so if we regard density as a quantity it must be the ratio of a mass to a volume.

Now it is often argued that multiplication and division have no meaning except in application to numbers. The contrary was vigorously stated by W. Stroud. The tendency of pure mathematics has long been to enlarge its scope and not to abandon expressions as meaningless unless they are shown to lead to actual inconsistencies. Stroud's views have succeeded to the extent that physical measures are usually stated in a way that explicitly contains products and ratios of physical magnitudes, but the grounds for them are not so well known as they should be. The essential point is that 'A gm. per cubic centimetre' is a linguistic expression, and there are two conditions that it ought to satisfy if it is to be useful. Are the conditions for its use clearly stated? And can it lead to inconsistencies if it is used in accordance with those conditions? On the first point, if A is the number associated with density when the volume is in cubic centimetres and the mass in grams, the density is taken to be $\rho = A$ gm./1 c.c.; that is all the definition needed. If we have another portion of liquid of volume $v' = x'$ cubic centimetres, and its mass is $m' = y'$ gm., then $y' = Ax'$ and

$$m' = y' \text{ gm.} = Ax' \text{ gm.} = Av' \text{ gm.}/1 \text{ c.c.} = \rho v'.$$

Thus the rule does provide a way of determining the mass of any portion of a liquid when the volume is given.

On the second point, it remains to be seen whether the interpretation of ρ is the same whatever the choice of units. Suppose the new unit of mass is μ gm. (call it a pound) and the new unit of volume is ϕ cubic centimetres (call it a pint). Then the mass of a pint of liquid is $A\phi$ gm $= A\phi/\mu$ pounds, and the density is $\dfrac{A\phi \text{ pounds}}{\mu \text{ pint}}$. But since 1 pound$/\mu =$ 1 gm. and 1 pint$/\phi =$ 1 cubic centimetre, this is just A gm./1 cubic centimetre, which was our original value of ρ. Hence the use of the language of quantity for density can lead to no inconsistency. In view of its general usefulness, it can therefore be strongly recommended.

5·5. *Dimensions.* Any quantity capable of additive measurement can be considered to have its own appropriate type of dimension. Since merely multiplying, say, a length by a number gives a length, number is taken to have zero dimension. However, it is often natural to use relations between additive magnitudes in such a way as to express the dimensions of some in terms of others, usually by replacing a derived magnitude by a number, usually 1. In the above discussion of density, for instance, volume has been taken to have its own dimension; as an additive magnitude it would be measured by the number of times the portion of fluid considered fills a standard vessel. But the volume of a vessel in the form of a rectangular parallelepiped, in this sense, is found to be proportional to the product of the sides, which is the cube of a length. The ratio is a derived magnitude, which we put equal to 1. Then the volume can be computed directly from the linear dimensions. If the unit of length is 1 cm., the unit of volume will be the volume of a centimetre cube, and volume has dimensions (length)³. Units connected by taking numerical factors equal to 1 are usually called *consistent*. This term is not very satisfactory because it is quite consistent to measure length in inches and volume in pints; the word *germane*, suggested by E. A. Guggenheim, is better.

It is possible to express the dimensions of all the quantities of mechanics in terms of three, length, time and mass, as follows:

length, L	time, T	mass, M
velocity, L/T	acceleration, L/T^2	force, ML/T^2
displacement, L	gradient of displacement, o	
stress = force per unit area, M/LT^2		density, M/L^3

elasticity = stress per unit gradient of displacement, M/LT^2
gradient of velocity, $1/T$
viscosity = stress per unit velocity gradient, M/LT
square of elastic wave velocity = elasticity/density, L^2/T^2
kinematic viscosity = viscosity/density, L^2/T.

Having the dimensions we can write down the germane units at once. For instance, if the units of length, time and mass are a centimetre, a second, and a gram, the germane unit of stress is $1 \, \text{gm.}/\text{cm.}\,\text{sec.}^2$.

It is widely supposed that dimensions are concerned entirely with transformations of units. This is not so. Dimensions of additive magnitudes arise through the method of measurement itself; and even if we never had to change units the dimensions of a derived magnitude arise in describing the property that it measures. Dimensions do help in transformation of units, but dimensions come first. One important feature is that quantities of the same dimensions can often be added, either in experiment or in calculation; quantities of different dimensions never can. This often provides a valuable check in theoretical work.

5·6. The pendulum. This affords a simple example of methods used to find the form of a solution without actually proceeding to the detailed investigation. We first take the method of dynamical similarity. The equation of motion is taken as

$$l\ddot{\theta} = -g \sin \theta, \tag{1}$$

and it is supposed known from experiment that the motion is periodic with period T. Take a new variable

$$t_0 = \sqrt{(g/l)} \, t. \tag{2}$$

Then the differential equation reduces to

$$\frac{d^2\theta}{dt_0^2} = -\sin \theta; \tag{3}$$

and t_0 is a number, since g is an acceleration and l a length. (This is necessary for consistency since θ is measured in circular measure and is therefore a number.)

Now pendulums may be of different lengths; they may also be in places where the values of g are different. But in any case this transformation can be carried out, and if there is a periodic solution it is

$$\theta = \theta(t_0) = \theta\{\sqrt{(g/l)} \, t\}. \tag{4}$$

Then if θ goes through the same cycle of values the time taken must be proportional to $\sqrt{(l/g)}$. Note that the last condition shows that the answer depends on θ attaining the same values and therefore on the amplitude being the same; in fact the period does depend also on the amplitude. Note also that the method is applicable only when the law is known. In this case the solution of the differential equation (3) has been calculated for all amplitudes, but if this was not so, it would remain possible to find the period experimentally for different amplitudes, and then generalization to all values of g and l would still be possible.

The method of dimensions does not assume the law known. It simply appeals to the fact that the physical parameters in the problem are the mass, the length, and gravity. If there is a period T in time, we assume a relation of the form

$$T = Nm^{\alpha}l^{\beta}g^{\gamma}, \qquad (5)$$

where N, α, β, γ are numbers. Writing the dimensions of the factors we have

$$T = M^{\alpha}L^{\beta}(L/T^2)^{\gamma}, \qquad (6)$$

whence for consistency

$$\alpha = 0, \quad \beta + \gamma = 0, \quad 1 = -2\gamma, \qquad (7)$$

and

$$\beta = \tfrac{1}{2}, \quad \gamma = -\tfrac{1}{2}, \qquad (8)$$

$$T = N\sqrt{(l/g)}. \qquad (9)$$

The drawback of this method is that it says nothing about the number N. The method of dynamical similarity indicates that N may depend on the amplitude, and in fact it does. But the method of dimensions gives no clue to what N may depend on. Further, in this case we know (1), and therefore that only l and g are relevant. But the dimensional argument is often applied where the law is not known, and then there is no guarantee that no additional non-dimensional parameters are relevant.

Another method, also often called the method of dimensions, would be better called the method of transformation of units. Suppose we multiply the units of mass, length and time by μ, λ, τ. Then the unit of g will be multiplied by λ/τ^2 (this assumes that we retain germane units, but there is no obligation to do so). The numbers associated with the measures will be divided by the same ratios. Then the numbers in the equation involving numbers only that corresponds to (5) will in general be altered, but the equation will remain true provided that

$$\tau = \mu^{\alpha}\lambda^{\beta}(\lambda/\tau^2)^{\gamma}. \qquad (10)$$

If this is to be true for all values of μ, λ, τ, we have equations (7), (8) again. But this tells us nothing new. We are still speaking of the original system, and (10) is simply a check on whether the units are germane. It tells us nothing whatever about a new system.

5·7. *Electricity and magnetism.* Electric charge can be measured in terms of a standard, by measuring the force on a given charge at a given distance. If the second charge is at B, the forces due to superposing charges at A are additive; thus charge is an additive magnitude. There is, however, a complication since the force is on another charge at B and therefore depends on both charges. All we can say at once is that the force is of the form

$$F_1 = \kappa_1 Q Q' / r^2, \tag{1}$$

where κ_1 is a constant, and Q, Q' are the charges. Thus $\kappa_1 Q^2$ is of dimensions ML^3/T^2.

The force between two magnetic poles is

$$F_2 = \kappa_2 \mathfrak{m} \mathfrak{m}' / r^2 \tag{2}$$

where $\mathfrak{m}$, $\mathfrak{m}'$ are the pole strengths and κ_2 is another constant. (It is theoretically more satisfactory to state the law in terms of magnetic moments, but this would only introduce extra complications here. For similar reasons we consider only forces in a vacuum.) Then $\kappa_2 \mathfrak{m}^2$ is of dimensions ML^3/T^2.

An electric current and a magnetic pole interact, with a force equal to that due to a magnetic shell occupying the circuit and with an intensity proportional to the current. This intensity is a magnetic moment per unit area and has dimensions $\mathfrak{m}/L$. We write the current as I and the equivalent strength of the magnetic shell as $\kappa_3 I$. Now an electric current is the amount of charge passing per unit time and therefore has dimensions Q/T, and the magnetic shell intensity has the dimensions of $\kappa_3 Q/T$. Thus $\kappa_3 Q/T$ and $\mathfrak{m}/L$ have the same dimensions.

Comparison of these relations shows that $\mathfrak{m}^2/Q^2$ has the dimensions of κ_1/κ_2, and therefore that $\kappa_1/\kappa_2 \kappa_3^2$ has dimensions L^2/T^2; that is, it is the square of a velocity. Hence we may be able without inconsistency to take two of κ_1, κ_2, κ_3 to be numbers; but to take all three to be numbers would be contradictory.

There are several systems of c.g.s. units. In the mixed or Gaussian system κ_1 and κ_2 are taken to be the number 1. Then for consistency $1/\kappa_3$ must be a velocity, and the dimensions of Q and $\mathfrak{m}$ are both $M^{\frac{1}{2}}L^{\frac{3}{2}}/T$. The germane unit of charge (e.s.u.) is such that two unit charges 1 cm. apart repel with a force of 1 dyne; and the germane unit of magnetic pole strength is such that two unit magnetic poles 1 cm. apart repel with a force of 1 dyne.

In the electromagnetic system (e.m.u.) κ_2 and κ_3 are taken to be 1. Then the dimensions of $\mathfrak{m}$ are $M^{\frac{1}{2}}L^{\frac{3}{2}}/T$ as before, and those of Q are $M^{\frac{1}{2}}L^{\frac{1}{2}}$. κ_1 is of dimensions L^2/T^2. The germane unit of magnetic pole strength is as before; unit electric current (e.m.u.) is defined so that unit current in a circuit produces the same magnetic field as unit magnetic moment per unit area over a surface occupying the circuit.

In the two systems the dimensions of electric charge differ by L/T. In most theoretical books, with a few exceptions (notably Sommerfeld's *Electrodynamics*, 1952), this astonishing result is accepted. One explanation offered is that the charges are defined through different phenomena, and there is no reason why they should have the same dimensions. This, however, overlooks the fact that they are additive magnitudes and can be compared directly. It is possible to charge a condenser by electrostatic methods and measure its charge electrostatically; and then to discharge it through a ballistic galvanometer and measure its charge electromagnetically. We thus get a direct comparison of the units as a number.

Another explanation is that, though the charges are comparable, there is no reason why a quantity must be measured in terms of a unit of the same dimensions as itself. This overlooks the fact that the definitions define actual charges; there is no question of their defining something different from what they set out to define.

The real reason for the inconsistency is that the comparison involves two sets of conventions, which are inconsistent. We saw in regard to the definition of volume that there is an element of convention in taking it to have dimensions L^3; volume has a natural method of additive measurement and can be given a dimension in its own right. The identification of this dimension with L^3 is convenient and does not lead to inconsistencies. But we have to be careful about conventions. By convention in English we call a ripe tomato red and grass green; it would be quite consistent to call a tomato green and grass red; but if we used both conventions at once we should infer that a tomato is the same colour as grass, which is false in either language. The point is that the comparison of units has assumed that κ_1, κ_2, κ_3 are numbers; but whatever dimensions are given to Q and $\mathfrak{m}$, $\kappa_1/\kappa_2\kappa_3^2$ has dimensions L^2/T^2 and cannot possibly be a number.

By experiment the e.m.u. of charge is found to be 3×10^{10} times the e.s.u. This ratio is not the velocity of light, as is often said; it is the ratio of the velocity of light to the germane unit of velocity, 1 cm./sec. The point is that if, for instance, we take the Gaussian definition as fundamental, $1/\kappa_3$ must be a velocity. In the electromagnetic system $\kappa_3 = 1$, and we can restore dimensional consistency without changing the numerical measures by replacing it by $\kappa_3 = 1$ sec./cm. Thus the experimental value of the ratio of the two units of charge shows that on the Gaussian system $1/\kappa_3$ would be 3×10^{10} cm./sec.

Campbell called length, time, and mass basic magnitudes, the important point being that with suitable conventions the rest of the mechanical magnitudes can be defined in terms of them. In both the Gaussian and the e.m. systems this is extended to the electric and magnetic quantities, but at the cost of introducing fractional dimensions. We are not compelled to adopt either system, and an alternative method is to take electric charge as an additional basic magnitude Q. Then the dimensions of κ_1 are ML^3/T^2Q^2. Since magnetic fields are largely, and perhaps entirely, due to the motion of charged particles there seems to be no need for a separate theory of magnetism, and we can take $\kappa_3 = 1$; then the dimensions of κ_2 are ML/Q^2. The dimensions of the principal quantities on the two systems are as in the following table:

	Gaussian	Charge
electric charge	$M^{\frac{1}{2}}L^{\frac{3}{2}}T^{-1}$	Q
magnetic pole	$M^{\frac{1}{2}}L^{\frac{3}{2}}T^{-1}$	LQT^{-1}
magnetic moment	$M^{\frac{1}{2}}L^{\frac{5}{2}}T^{-1}$	L^2QT^{-1}
force per unit charge	$M^{\frac{1}{2}}L^{-\frac{1}{2}}T^{-1}$	$MLT^{-2}Q^{-1}$
force per unit pole	$M^{\frac{1}{2}}L^{-\frac{1}{2}}T^{-1}$	$MT^{-1}Q^{-1}$
electric current	$M^{\frac{1}{2}}L^{\frac{3}{2}}T^{-2}$	QT^{-1}

When a law is used to express an additive magnitude in terms of others, Campbell called the resulting expression a quasi-derived magnitude. Thus volume, when taken as of dimensions L^3, is a quasi-derived magnitude; so are electric charge and magnetic pole strength in the first of the above systems, magnetic pole strength in the second.

There was a long discussion on this subject in 1940–3 in the *Proceedings of the Physical Society* and the *Philosophical Magazine*, to which many physicists contributed. Most supported one or other of the above systems, but nobody advocated having different dimensions for the electrostatic and electromagnetic units.

5·8. *Temperature.* This is defined originally in terms of a linear scale. A mercury thermometer bears marks at the places reached by a liquid at the freezing and boiling points of water, which are labelled 0° and 100°. The scale is graduated linearly. Thus a temperature as read on a thermometer is the ratio of two lengths multiplied by 100. It is therefore a number. A thermometer based on the expansion of a gas can be used more generally; with some small corrections the mercury thermometer can be adapted to the gas scale, and absolute temperature is obtained by adding 273.

Statistical mechanics provides an explanation of thermal phenomena, including thermal expansion, and it is found that the volume of a gas at given pressure should be proportional to the energy per degree of freedom. Thus it is possible also to regard absolute temperature as an energy. The thermometer reading is 100 times the ratio of two lengths, and could equally well be regarded as 100 times the ratio of two energies.

On the other hand all definitions in terms of expansion or energy become unsatisfactory at very low temperatures, and a definition in terms of thermodynamic efficiency is the only satisfactory one. This gives only ratios and has to be standardized in comparison with a gas thermometer at some standard temperature. This leaves the dimensions of the unit arbitrary.

CHAPTER VI

MENSURATION

'Tis distance lends enchantment to the view.
THOMAS CAMPBELL, *The Pleasures of Hope*

6·0. The most fundamental physical science is that of the relations between lengths. We shall call it *mensuration*.† It deserves a fairly full discussion because the usual methods of treatment confuse it with the subject known to pure mathematicians as *geometry*. The chief object of the following discussion is to bring into prominence the experimental basis, which is neglected in the usual treatment.

Geometry proceeds by taking some general axioms, irrespective of whether these are physically tested or capable of being tested, and develops their consequences by purely logical rules. Physical measurement plays no part in it. For us, physical measurement is the whole *raison d'être* of the subject. By comparing our measurements we reach certain generalizations, which in some respects resemble the axioms of some forms of geometry. But the structure is essentially different. Geometry starts with the laws, and the particular results are consequences of the laws. In mensuration the particular results are the essence of the matter, and the general laws are derived from them by a process of generalization based on the simplicity postulate.

It might appear that, in spite of the opposite modes of approach, the total contents of mensuration and geometry might be the same, the axioms of geometry being the same as the laws of mensuration. But this is not the case.

All projective and descriptive geometries can evidently be ruled out at once. A requirement of all such geometries is that no notion analogous to distance is to be used. Since distance forms our subject-matter, there is no common ground whatever.

Euclid's geometry‡ is the closest existing analogue of mensuration,

† The name was used in this sense by G. Atwood (of Atwood's machine) in *Lectures in Natural Philosophy* (1784), 192.

‡ The fullest account is the 3-volume work of Sir T. L. Heath, which contains also the algebraic books. The best I know of ordinary size is Nixon's *Euclid Revised*.

and requires a full discussion. The notion of length is freely used in it. His points are sufficiently like what we have so far called marks or the ends of objects, and we can produce close physical analogues of his straight lines, planes, and circles. He freely uses the principle of juxtaposition as a test of whether one quantity is greater than another; and this is essential to the physical method of comparing lengths.

Nevertheless his system differs from any possible account of mensuration; in fact it is neither a mensuration nor a geometry, but a mixture of the two. For instance, he uses compasses to draw circles, a legitimate physical procedure, but refuses to use them to transfer distances. In I (2), when he wishes to draw from a given point a line whose length is equal to that of a given straight line not through the point, he makes a complicated construction to avoid having to lift up the compasses and transport them. Yet in I (4), in testing the equality of two triangles, he supposes one picked up bodily and superposed on the other. The ordinary properties of rigid bodies are supposed to be possessed by triangles (drawn on pieces of paper) but not by a pair of compasses. The usual criticism from the geometrical standpoint is to reject the proof of I (4) and provide a new one; from the physical standpoint the proof of I (4) is valid in certain conditions, though the result is true even when the construction involved cannot be carried out, but the complication of I (2) avoids only a difficulty that has already been dealt with in giving a meaning to length.

Euclid postulates further that any two points can be joined by a straight line. The physical analogue of this is often true, but not always. The points may be on the surface of a convex solid too hard to be bored. Yet the distance between the points exists, for it can be measured by applying compasses first to the two points and then to a scale. Physically the notion of the distance between two points is more general than that of the straight line joining them.

The most important departure of Euclid's treatment from any possible account of mensuration, however, is in the discussion of parallels and the related propositions. We may refer to the second postulate, that a straight line can be produced to any length, however great, and to the fifth postulate, also called the twelfth or parallel axiom. Both of these postulates have been criticized by modern geometers as not obvious. In mensuration, on the other hand, they not only are not obvious but are false. We cannot produce a physical straight line to a length greater than one determined by the size of the body it is drawn on. Again, it may be possible to find out by our existing methods of measuring angles that, when one straight line crosses two others, it makes the sum of the interior angles less than two right angles, but it does not happen in all such cases that the two straight edges it crosses intersect; for in practice they often cannot be made long enough, or they may not be in one plane—a detail not allowed for in the usual statement of the postulate.

The alternative known as Playfair's axiom does not meet the difficulty, for it is not true that of two intersecting straight lines one at least must intersect any other; Playfair's parallel axiom fails in just the same way as Euclid's.

Criticisms of Euclid's *Elements* have usually been made from the geometrical standpoint and not from the physical one, and its virtues from one are usually its vices from the other. His test of equality is always superposition. Two lengths are equal if one can be superposed on the other. The same applies to two angles. Two areas are equal if one can be cut up and the pieces placed so that they exactly cover the other. These are physical methods of comparison; and just for that reason they are rejected by geometry. His procedure is such that numerical measures do not arise; addition and subtraction are done on the actual objects themselves. He therefore sacrifices the convenience of being able to resort to algebra; but he also avoids a trap. Euclid would never have said that a length of $1\cdot5$ cm is converted into one of $1\cdot5$ in. by a 'mere change of units'; nor would he have said that the mass of the Sun is $1\cdot5$ km.

The word 'geometry' literally means the measurement of the Earth, and Euclid's predecessors were largely inspired by the needs of surveying. By this time, however, the name has become so closely connected with the branch of pure mathematics that it seems hopeless to attempt to rescue it. Nor is it, I think, worth while. The measurement of the Earth is now generally known as geodesy, and what we need is a word to describe the theory of measurement of length in general, not merely in relation to the Earth. 'Mensuration' seems entirely satisfactory, saying neither more nor less than it is intended to mean.

It is a fact that when Euclid's theory gives a quantitative answer and the relevant construction can be carried out, the result is always found to be physically correct.† Nevertheless his axioms assume so many things possible that are in fact physically impossible that a radical reconstruction is needed. The task of reconstructing Euclidean mensuration is difficult. An extended discussion of substitutes for the parallel axiom was given by C. L. Dodgson (otherwise Lewis Carroll) in one of his less read works.‡ The main conclusion was that, though many substitutes could be shown equivalent to the parallel axiom, none was more suitable for introduction at an elementary stage. A reconstruction as a geometry, with full statements of the axioms used by Euclid without explicit statement, is by H. G. Forder.§

The modern physicist will not share Euclid's antipathy to measurement, and will recognize in his treatment of angles and areas that these, like length, are additive magnitudes. If he accepts the notion of quantity he will not refuse to say that a square centimetre is literally the square of a centimetre, but it is not strictly necessary to say so. The question that does arise at the outset is whether we should introduce from the start any additive magnitudes besides length.

Euclid assumes in I (13) that if a pencil of coplanar lines is drawn through a point, the angle between the extreme lines is equal to the sum of those between consecutive lines of the pencil. In I (4) he supposes that angles that can be superposed are equal. These postulates make it theoretically possible to construct a scale for measuring angles in terms of a unit. Such a scale we may at once call a *protractor*, and angle is an additive magnitude. Again, in I (35) he compares the areas of parallelograms by cutting them up and superposing them, and his later work with triangles and rectangles indicates that area also is treated as an additive magnitude. There are therefore three different additive magnitudes in the theory, and in the development they continually influence one another. All can be shown to exist in the sense that their measurement can actually be carried out, and there is no theoretical objection

† Geodesy possibly provides the most sensitive tests, distant base-lines being compared by trigonometric methods. There is slight disagreement in the extreme case of the displacement of star images by the Sun's gravitational field.

‡ *Euclid and his Modern Rivals* (1879).

§ *The Foundations of Euclidean Geometry* (1927).

to developing the theory of all together. But there is a practical objection. Angles and areas can actually be superposed only in special cases; projections on the bodies that carry them usually interfere with the superposition. Again, the addition of angles depends on their being in the same plane; thus the direct measurement of angle depends on the existence of physical planes, whereas the measurement of distances by means of compasses and scale is independent of the existence of planes. Since distance is much more generally measurable than either angle or area it is desirable to develop the theory, as far as possible, on the basis of the properties of distance alone.

6·1. Mensuration deals essentially with the relations between measurements of distance, and in the first place these measurements are on rigid bodies. It may be suggested that, before it can be discussed, we should define the terms 'distance' and 'rigid bodies'. Now the requirement of a definition is that it must make it possible to recognize the defined entity when it actually occurs. It is of no value to us to say that a rigid body is one such that the distances between all pairs of points on it are unaltered by any displacement, nor to define relative motion as change of distance between parts of a system, unless we have some way of recognizing when distances *are* altered. Distance, again, cannot be defined in terms of the properties of rigid bodies unless we have first some way of recognizing the rigid body when we meet it. None of these notions can be defined in terms of the properties of 'space', because we have no means of recognizing space directly; distance in space, for instance, cannot be determined except through measurements, which at once reintroduce material scales, which the reference to space was intended to avoid.

The solution seems to be that neither the rigid body nor distance is directly recognizable, and that both are derived from still more elementary notions, several experimental facts being used in the process. We can make permanent marks on bodies, which we can recognize afterwards. By means of compasses or calipers we can compare pairs of marks with one another. All pairs of marks that can be fitted by the compasses in the same adjustment can be classified together. We abstract the common property of *distance* and say that all such pairs are equidistant. Now when a fit of pairs of marks on the same body has been obtained, it may be found that a fit is always obtained again in subsequent trials. If this holds for numerous pairs of marks on the same body, we can generalize it as a law for that body. Such a body is called *rigid*. Compasses are rigid bodies provided their adjustment is not altered. If there is a doubt as to whether their adjustment has altered, they can be tested by application to several pairs of marks that they previously fitted, and if they fail we can tighten up the hinge or get a new pair. In the first place distance is simply a property of pairs of marks on rigid bodies.

So far we have defined only identity of distance and not greater and less in relation to distance; it is enough to say that the distance from A to B is less than the distance from C to D if the compasses, having been set to fit A, B, have to be set to a wider adjustment to fit C, D.

Measurement of distance depends on the existence of bodies with straight edges. If two bodies touch at two points they may touch at a continuous set of intermediate points. In general, when this is done, if we turn one or both of the bodies about so that they remain in contact at two different pairs of marks, the intermediate points that were formerly in contact separate. But it is again a fact that bodies can be made with such edges that they do remain in contact when they are turned about two coincident marks. When this has been found to hold in a large number of trials we can infer with a high probability that it will hold in any subsequent trial. In such cases we call the edges *straight*. This is the analogue of Euclid's axiom that two straight lines cannot enclose a space.

We can now proceed to the construction of a measuring scale on a straight edge and to the actual measurement of distances by the principles of the previous chapter.

The reservation must be made that the bodies must receive only ordinary treatment. It is easy to recognize by what we call sensations of force when exceptional treatment is taking place. If bodies or edges fail to satisfy our tests we say that they are not rigid or not straight, or that exceptional treatment has taken place. In that case they do not form part of our present subject-matter. The important thing is that there are many bodies that do satisfy the conditions. If all compasses were made of rubber and all bodies of plasticine, this would not be so, and perhaps there would be no science of mensuration; but actually we can classify bodies and edges according as they do or do not satisfy the tests, and confine our attention for the present to those that do. The others are reserved for the subjects of mechanics.

6·2. So long as marks lie on the same straight edge, the distances between them follow the simple rules of addition and subtraction. But we also require propositions connecting distances between marks not on the same straight edge. This brings us into a new domain, and at least one new experimental fact is needed to serve as a starting point. As has already been indicated, propositions involving angles or planes should be avoided until we can interpret them in terms of lengths. Our physical treatment must begin with an experimental law connecting distances between marks not on the same straight edge.

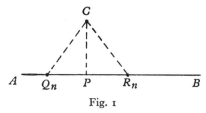

Fig. 1

6·21. If P is a mark on a straight edge AB (Fig. 1), not at either end, we can use the compasses to find two points of AB equidistant from P and on opposite sides of it. Now if C is not on AB, but sufficiently near some point of AB, we can similarly use the compasses to find pairs of points, say Q_n, R_n equidistant from C. The mid-point of $Q_n R_n$ is always the same point. We shall call this P, the foot of the perpendicular from C to AB.†

It is then found that

$$CQ_n^2 = CR_n^2 = CP^2 + PQ_n^2 = CP^2 + PR_n^2.$$

This property, that the square of one side is the sum of the squares of the other two, will be our definition of a right-angled triangle.

6·22. Now consider two straight edges OA, OB meeting in O (Fig. 2), and consider a mark X on OA. Take a length L and consider

† In the figures continuous lines denote actual straight edges; dotted lines connect marks the distances between which are considered, but which need not be connected by actual straight edges.

points Y on OB such that $XY = L$. There are either $0, 1$ or 2 positions of Y such that this is true. This suggests that, given XY, the possible values of OY are the roots of a quadratic equation

$$OY^2 - 2b.OY + c = 0 \qquad (1)$$

that lie between 0 and OB; b and c may depend on OX and XY. The mean of the roots is b. But we have seen that for given X the mean of the values of OY consistent with XY being given is independent of XY; hence b is independent of XY.

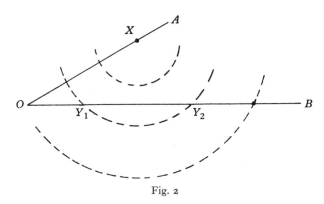

Fig. 2

Again, taking X' and Y' on OA and OB, such that $OX' = OY$, $OY' = OX$, we find that $X'Y' = XY$. Hence the left side of (1) must be symmetrical in OX and OY, and therefore must have the form

$$OX^2 + OY^2 - 2\lambda OX.OY - f(XY), \qquad (2)$$

where λ is a constant. The function f is so far unspecified, but if we take $XY = OX$ one position of Y is at O and then $OY = 0$. Then

$$OX^2 - f(OX) = 0. \qquad (3)$$

This holds for all positions of X on OA and therefore the function is identified; and finally for all pairs X on OA and Y on OB

$$XY^2 = OX^2 + OY^2 - 2\lambda OX.OY. \qquad (4)$$

The relation can also be written:

$$\lambda = \frac{OX^2 + OY^2 - XY^2}{2OX.OY} \qquad (5)$$

is independent of OX and OY. In the case where AOB is straight, $XY = OX + OY$ and then $\lambda = -1$. If OX passes through Y or OY passes through X, $\lambda = 1$.

So far the argument leading to (5) is suggestive; it depends on the assumption in (1) that the possible values of OY are the roots of a quadratic equation, but there are functional equations that have at most two real roots without being quadratic. But (5) is in a form that can be tested by direct measurement. Such a test has already been carried out in countless experiments in practical plane 'geometry'. It has perhaps not been tested directly with the full accuracy of modern measuring apparatus, but many of its consequences have. We can therefore adopt (5) as a fundamental law connecting distances between three points on two intersecting lines.

6·23. In trigonometry λ is the cosine of the angle between OA and OB. We have not yet, however, defined 'angle' quantitatively. What we can do is to note that (5) is a property of the pair of edges OA, OB, independent of the selection of X and Y on them, and we can give it the name $\cos AOB$.

If a third straight edge OC passes through O, it may be possible to place another straight edge so as to meet all of OA, OB, OC. Let the intersections be X, Y, Z. Then if Y is between X and Z,

$$\cos XOY > \cos XOZ, \tag{1}$$

that is, $$\cos AOB > \cos AOC. \tag{2}$$

The values of λ can thus be ordered according to the intersections of a transversal with edges, where such intersections exist or can be constructed. It would be convenient in such a case to have a way of specifying a numerical value α associated with each value of λ so that in such a case

$$\alpha(AOC) = \alpha(AOB) + \alpha(BOC). \tag{3}$$

Then α will have an additive property. This would be achieved by a protractor; alternatively we can use the analytic definition of the cosine. This is

$$\cos x = 1 - \frac{x^2}{2!} + \frac{x^4}{4!} - \frac{x^6}{6!} + \dots. \tag{4}$$

We take also the companion functions

$$\sin x = x - \frac{x^3}{3!} + \frac{x^5}{5!} - \frac{x^7}{7!} + \dots \qquad (5)$$

$$\tan x = \sin x / \cos x. \qquad (6)$$

The chief properties of these functions for our present purpose are:

(1) $\cos 0 = 1$, $\sin 0 = 0$.

(2) $\cos x$ decreases as x increases, and takes the value 0 when $x = \frac{1}{2}\pi$ and -1 when $x = \pi$. Thus for every value of $\cos x$ in $1 \geqslant \cos x \geqslant -1$ there is one value of x in $0 \leqslant x \leqslant \pi$, and conversely.

(3) $\cos^2 x + \sin^2 x = 1$.

(4) $\sin x$ increases with x till $x = \frac{1}{2}\pi$, when $\sin x = 1$; it then decreases again and vanishes at $x = \pi$.

(5) $\cos(x_1 + x_2) = \cos x_1 \cos x_2 - \sin x_1 \sin x_2$,

$\sin(x_1 + x_2) = \sin x_1 \cos x_2 + \cos x_1 \sin x_2$.

These statements are essentially from pure mathematics and do not depend on the definitions of the cosine and sine in trigonometry. The point is that writing $\lambda = \cos \alpha$ specifies a number α associated with two intersecting edges, which can range from 0 to π and will later be identified with the circular measure of the angle between them. We have not proved that α so specified has the additive property. If $\cos XOY = \lambda$ we shall write the value of α corresponding to it as $\angle XOY$ (read, angle XOY).

In this form, 6·22 (4) is practically Euclid's II (12) and (13). In 6·21, (4) shows that the angle at P is $\frac{1}{2}\pi$.

6·24. It may happen that for three marks O, A, B not on a straight edge (i.e. not with any material connexion between them) we can measure the three distances between them and evaluate λ. If $\lambda = \pm 1$ we call the marks *collinear*. However, if λ is near ± 1 the difference from ± 1 depends on a difference that is small compared with the quantities subtracted, and thus α is not accurately determinable by measurement of distances alone. But in this case it is possible to look along the line from, say, O to B, and if B is then behind A or in front of it we can again call the marks collinear. This use of the rectilinear propagation of light is in these conditions much more sensitive to the value of α than measurement of distances is. Incidentally it brings the telescope into the subject of mensuration.

It can now be found that if O, X_1, X_2 are collinear, and O, Y_1, Y_2 are collinear,

$$\frac{OX_1^2 + OY_1^2 - X_1Y_1^2}{2OX_1.OY_1} = \frac{OX_2^2 + OY_2^2 - X_2Y_2^2}{2OX_2.OY_2}$$

irrespective of whether the marks are on actual straight edges. Also in experiments on a laboratory scale, if O and X are assigned, we can place Y in precisely one way so that O, X and Y are collinear and OY has a given value.

We can now proceed to develop the theory.

It may be found rather dull—of the few that read it, perhaps most would use a stronger adverb. There is nothing new in the data or in the results. But the method is not given elsewhere, at any rate not as a whole, and I wish to emphasize, in the most elementary physical subject, some features of the method. These features also are not new, but have usually been thought to be something to be ashamed of, and therefore to be disguised as far as possible. My object is to bring them into the daylight and to explain why there is no need to be ashamed of them.

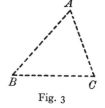

Fig. 3

6·31. If A, B, C are three marks as in Fig. 3 we have

$$\cos BAC = \frac{AB^2 + AC^2 - BC^2}{2AB.AC}. \tag{1}$$

Write　　　　　　$BC = a,\ CA = b,\ AB = c,\ a+b+c = 2s.$
Then

$$\cos BAC = \frac{b^2 + c^2 - a^2}{2bc} \tag{2}$$

$$\sin \tfrac{1}{2}BAC = \left(\frac{1 - \cos BAC}{2}\right)^{\frac{1}{2}} = \left(\frac{-(b-c)^2 + a^2}{4bc}\right)^{\frac{1}{2}} = \left(\frac{(s-b)(s-c)}{bc}\right)^{\frac{1}{2}} \tag{3}$$

$$\cos \tfrac{1}{2}BAC = \left(\frac{1 + \cos BAC}{2}\right)^{\frac{1}{2}} = \left(\frac{(b+c)^2 - a^2}{4bc}\right)^{\frac{1}{2}} = \left(\frac{s(s-a)}{bc}\right)^{\frac{1}{2}} \tag{4}$$

$$\tan \tfrac{1}{2}BAC = \frac{\sin \tfrac{1}{2}BAC}{\cos \tfrac{1}{2}BAC} = \left(\frac{(s-b)(s-c)}{s(s-a)}\right)^{\frac{1}{2}} \tag{5}$$

Corresponding formulae for the other angles are obtained symmetrically. Note that since $-1 \leqslant \cos BAC \leqslant 1$, $\cos^2 \frac{1}{2} BAC \geqslant 0$; hence $s \geqslant a$, and therefore $b+c \geqslant a$; similarly $c+a \geqslant b$, $a+b \geqslant c$.

6·32. By the formula for the tangent of the sum of two angles

$$\tan \tfrac{1}{2}(BAC + BCA) = \frac{\tan \frac{1}{2} BAC + \tan \frac{1}{2} BCA}{1 - \tan \frac{1}{2} BAC \tan \frac{1}{2} BCA}$$

$$= \left(\frac{s(s-b)}{(s-c)(s-a)} \right)^{\frac{1}{2}} = \frac{1}{\tan \frac{1}{2} ABC}. \quad (6)$$

First take $\angle \frac{1}{2} ABC < \frac{1}{2}\pi$. Then in these intervals we can have (6) only if

$$\tfrac{1}{2}(\angle BAC + \angle BCA) + \tfrac{1}{2}\angle ABC = \tfrac{1}{2}\pi, \quad (7)$$

that is, $\qquad \angle BAC + \angle ABC + \angle BCA = \pi. \quad (8)$

If $\angle ABC = \pi$, A, B, C are collinear and $\angle BAC = \angle ACB = 0$. Then (8) is still true.

This can be read as equivalent to Euclid I (32), that the sum of the angles of a triangle is two right angles. But its meaning is different because Euclid's result depends on a method of physical addition of angles, whereas the angles here are defined in terms of the lengths of the sides, and (8) is in fact a mathematical identity for the distances between three points.

Note that the ratios of a, b, c determine the cosines and hence the angles uniquely; and conversely the angles determine the ratios of the sides uniquely, since

$$\frac{\sin BAC}{a} = 2 \frac{\{s(s-a)(s-b)(s-c)\}^{\frac{1}{2}}}{abc}, \quad (9)$$

which is a symmetrical function of the sides.

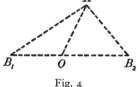

Fig. 4

6·33. If B_1, O, B_2 are collinear marks (Fig. 4) and A is another mark not collinear with them

$$AB_1^2 = OA^2 + OB_1^2 - 2OA \cdot OB_1 \cos AOB_1, \quad (1)$$

$$AB_2^2 = OA^2 + OB_2^2 - 2OA \cdot OB_2 \cos AOB_2, \quad (2)$$

and also $\qquad = AB_1^2 + B_1 B_2^2 - 2AB_1 \cdot B_1 B_2 \cos AB_1 B_2. \quad (3)$

But $$B_1B_2 = B_1O + OB_2. \tag{4}$$

$$\cos AB_1B_2 = \cos \angle AB_1O = \frac{AB_1^2 + OB_1^2 - AO^2}{2AB_1 . OB_1}$$

$$= \frac{OB_1 - OA \cos AOB_1}{AB_1} \tag{5}$$

by (1). Substituting from (4) and (5) in (3) we have

$$AB_2^2 = AB_1^2 + B_1B_2^2 - 2B_1B_2(OB_1 - OA \cos AOB_1)$$

$$= OA^2 + OB_1^2 + B_1B_2^2 - 2B_1B_2 . OB_1$$

$$+ 2OA(B_1B_2 - OB_1) \cos AOB_1$$

$$= OA^2 + OB_2^2 + 2OA . OB_2 \cos AOB_1, \tag{6}$$

whence, comparing (2) and (6), we have

$$\cos AOB_1 + \cos AOB_2 = 0, \tag{7}$$

and therefore $$\angle AOB_1 + \angle AOB_2 = \pi. \quad \text{(Euc. I (13).)} \tag{8}$$

6·34. It follows as an immediate corollary that when two straight edges intersect the opposite angles are equal (Euc. I (15)).

6·35. It also follows from 6·32 that in Fig. 4

$$\angle AB_1O + \angle B_1OA + \angle OAB_1 = \pi, \tag{1}$$

$$\angle OB_2A + \angle B_2AO + \angle AOB_2 = \pi, \tag{2}$$

$$\angle B_1B_2A + \angle B_2AB_1 + \angle AB_1B_2 = \pi. \tag{3}$$

Add (1) and (2) and subtract (3); after cancelling we have

$$\angle B_1OA + \angle AOB_2 + \angle OAB_1 + \angle B_2AO - \angle B_2AB_1 = \pi. \tag{4}$$

But by 6·33 $$\angle B_1OA + \angle AOB_2 = \pi. \tag{5}$$

Hence $$\angle OAB_1 + \angle B_2AO = \angle B_2AB_1. \tag{6}$$

Thus angles so placed that they have one common arm and so that collinear points exist on the arms, and measured by the rules given in 6·23, have the additive property. This identifies our measure of an angle with Euclid's, where the latter can be carried out, except for a constant factor.

6·36. If two straight edges OA and OB meet in O, and if $\cos AOB$ is negative, and if O is not the end of OB, it is possible to make a mark B_1 on OB so that $\cos AOB_1$ is positive.

For by 6·33 we need only take B_1 on the side of O opposite to B, and the result follows.

6·37. If A is outside the edge OB (Fig. 5), and if B is on that part of it where $\cos AOB$ is positive, and if OB is greater than $OA \cos AOB$, then it is possible to make a mark C on OB such that

$$OA^2 = OC^2 + AC^2.$$

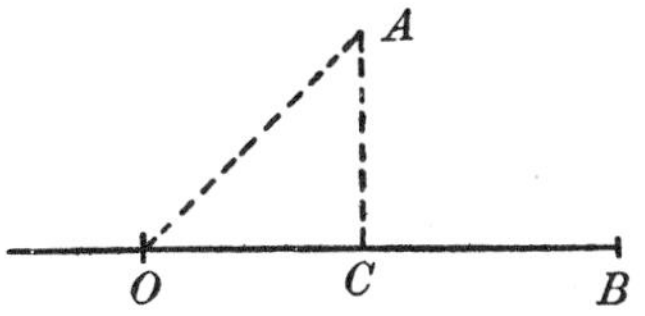

Fig. 5

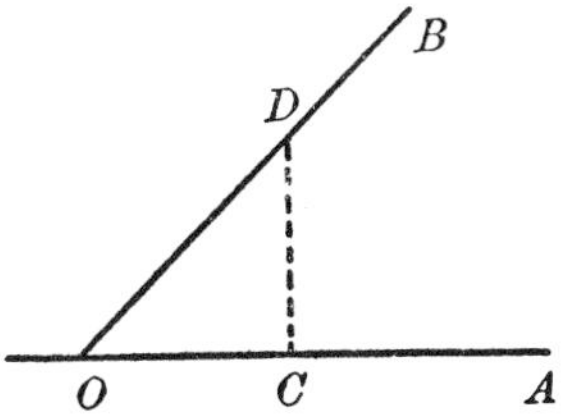

Fig. 6

For we can make C on OB at a distance $OA \cos AOB$ from O. Then

$$AC^2 = OA^2 + OC^2 - 2OA\,.\,OC \cos AOB = OA^2 - OC^2,$$

which proves the proposition. Also

$$\cos OCA = \frac{OC^2 + CA^2 - OA^2}{2OC\,.\,CA} = 0,$$

whence $$\angle OCA = \tfrac{1}{2}\pi.$$

We have therefore constructed a triangle with one angle equal to $\tfrac{1}{2}\pi$. (This is an alternative to the method used in 6·21.)

If the angle between two intersecting straight edges is $\tfrac{1}{2}\pi$, they are said to be perpendicular. C in the figure is called the foot of the perpendicular from A to OB.

6·38. If two straight edges OA, OB meet at O (Fig. 6) and C is any mark on OA, and if the length of OB exceeds $OC \sec AOB$, then we can make a mark D on OB such that C is the foot of the perpendicular from D to OA.

For we can make D such that $OD = OC \sec AOB$, and the result follows as in 6·37.

8

6·39. If the angle AOB (Fig. 7) is $\frac{1}{2}\pi$, we have

$$\cos OAB = \frac{OA^2 + AB^2 - BO^2}{2OA.AB} = \frac{OA}{AB}, \tag{1}$$

$$\sin OAB = \cos\left(\frac{1}{2}\pi - OAB\right) = \cos OBA = \frac{OB}{AB}, \tag{2}$$

$$\tan OAB = \frac{\sin OAB}{\cos OAB} = \frac{OB}{OA}, \tag{3}$$

with corresponding formulae for the other trigonometric functions. These results thus emerge as laws, and not as definitions of the functions as in ordinary trigonometry.

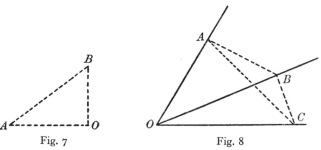

Fig. 7   Fig. 8

6·40. Consider three edges meeting in a point O (Fig. 8). It is always possible to fix A in one of them so that A is the foot of the perpendicular from marks B and C on the other two, since the condition of 6·38 can always be satisfied by taking OA short enough. Then

$$AB = OA \tan AOB, \tag{1}$$

$$OB = OA \sec AOB, \tag{2}$$

$$AC = OA \tan AOC, \tag{3}$$

$$OC = OA \sec AOC, \tag{4}$$

$$BC^2 = OB^2 + OC^2 - 2OB.OC \cos BOC$$
$$= OA^2(\sec^2 AOB + \sec^2 AOC - 2\sec AOB \sec AOC \cos BOC).$$

Also $\tag{5}$

$$BC^2 = AB^2 + AC^2 - 2AB.AC \cos BAC,$$
$$= OA^2(\tan^2 AOB + \tan^2 AOC - 2\tan AOB \tan AOC \cos BAC). \tag{6}$$

Equating (5) and (6), and multiplying by $\cos AOB \cos AOC$, we have

$$\cos BOC = \cos AOB \cos AOC + \sin AOB \sin AOC \cos BAC. \quad (7)$$

This formula introduces the third dimension for the first time, for it allows two different lines AB, AC to be both perpendicular to an edge OA at the same point. The formula (7) is the analogue of a familiar one in spherical trigonometry, though the sphere as such has not yet appeared.

It follows as a corollary that $\angle BAC$ is independent of OA.

If B, A, C are collinear, $\cos BAC = -1$, and (7) leads to

$$\angle BOC = \angle AOB + \angle AOC.$$

This is equivalent to §6·35 when actual straight edges connect the marks.

6·41. We can now proceed to the discussion of *planes*. If we take two fixed marks O and O', and any path from O to O', then at one end of the path the distance from O is greater than the distance from O', and at the other end the opposite is true. Both distances vary continuously, and therefore there is a possible position on the path such that the distances from O and O' are equal. Marks equidistant from two fixed marks are said to be *on the plane* determined by the fixed marks. We do not assume the construction of the complete plane (a perfectly flat surface is in fact much harder to make than a straight edge).

6·42. If two marks P and Q are on a plane (Fig. 9), every mark R collinear with them is also on the plane. For since P and Q are on the plane we can write

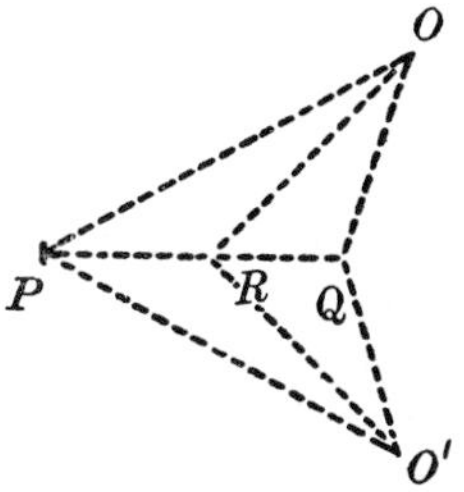

Fig. 9

$$PO = PO' = p; \quad QO = QO' = q; \quad PQ = r,$$

and $\quad \cos OPQ = \cos O'PQ = k$ say.

Then
$$RO^2 = PO^2 + PR^2 - 2PO \cdot PR \cos OPQ$$
$$= PO'^2 + PR^2 - 2PO' \cdot PR \cos O'PQ$$
$$= RO'^2.$$

6·43. If two planes are determined by pairs of marks O and O', H and H', three circumstances may arise. All positions P on the first plane may be equidistant from H and H'; then the planes are identical.

All positions in the first plane may be such that $PH > PH'$, or all such that $PH < PH'$; then the planes have no common point and may be said to be *parallel*.

Some positions in the first plane may be such that $PH > PH'$, and some such that $PH < PH'$. Then we may classify positions on the first plane according to the sign of $PH - PH'$. On any path from a position where this is positive to one where it is negative, the difference varies continuously and therefore passes through the value zero. Thus it is possible to assign marks common to the two planes; they are said to be *on the line of intersection of the two planes*.

Note that it is impossible for two planes to have just one point in common.

If Q and R are on the intersection of two planes, every point collinear with them is also on the intersection. Hence if there is an intersection it includes one straight line.

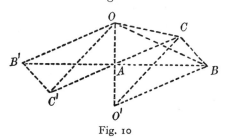

Fig. 10

6·44. Any three marks A, B, C lie on a plane. Take B' collinear with BA so that $B'A = AB$, and C' collinear with C, A so that $C'A = AC$. Then B and B' determine one plane and C and C' another. A lies on both planes. If O is another common point (Fig. 10), we have

$$OB' = OB; \quad OA = OA: \quad AB' = AB,$$

and therefore $\angle B'AO = \angle OAB = \tfrac{1}{2}\pi$.

Thus AO is perpendicular to AB, and similarly to AC. Now take O' collinear with OA so that $AO' = AO$. Then

$$\angle OAB = \angle O'AB = \tfrac{1}{2}\pi; \quad AO = AO'; \quad AB = AB;$$

and therefore $\qquad\qquad O'B = OB,$

and similarly $\qquad\qquad O'C = OC.$

Thus A, B, C are all on the plane determined by O and O'.

It follows that if two planes intersect in a straight line and have a common point not in this line, the planes are identical. For if A, B are in the line and C is a common point not on the line, A, B, C determine the plane completely. Conversely, if two planes are neither identical nor parallel, all their common points are collinear.

At this point we use the principle that there are not more than three dimensions. If there were four dimensions there might be infinitely many lines OO' through A, all perpendicular to AB and AC. The assumption that there is only one such line ensures that the points equidistant from O and O' are identical with those for which there are collinear points, other than A, in AB and AC.

6·45. In general a line has one point in common with a plane. For if P and Q are marks on the line, of which neither is on the plane (Fig. 11) and O, O' determine the plane, and R is another point on the line, in the direction PQ,

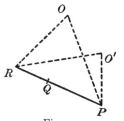

Fig. 11

$$OR^2 = OP^2 + PR^2 - 2OP.PR\cos OPQ, \quad (1)$$

$$O'R^2 = O'P^2 + PR^2 - 2O'P.PR\cos O'PQ, \quad (2)$$

and therefore $OR = OR'$ if

$$2PR(O'P\cos O'PQ - OP\cos OPQ) = O'P^2 - OP^2. \quad (3)$$

$O'P^2 - OP^2 \neq 0$; if then

$$O'P\cos O'PQ - OP\cos OPQ \quad \text{and} \quad O'P^2 - OP^2$$

have the same sign, there is a positive value of PR satisfying (3). If they have opposite signs there is no positive value of PR satisfying (3), but there is a suitable point R in the direction opposite to PQ. Thus unless $O'P\cos O'PQ = OP\cos OPQ$ there is precisely one point R satisfying the conditions. In the exceptional case no R satisfies the conditions; in this case we say that the line is *parallel* to the plane.

6·46. It follows that in general there is one point common to three planes.

6·47. Lines in a plane are said to be *parallel* if they make the same angle, similarly situated, with a given line.

Parallel lines have no common point at a finite distance. For if (Fig. 12)

$$\angle CAB = \alpha = \angle DBE,$$

$$\angle DBA = \pi - \alpha.$$

If then AC and BD had a common point L, we should have the sum of the angles of the triangle ABL equal to

$$\alpha + (\pi - \alpha) + \angle ALB = \pi + \angle ALB.$$

But this is impossible since A, B, L are not collinear and $\angle ALB \neq 0$.

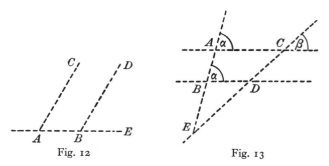

Fig. 12   Fig. 13

6·48. Any transversal intersecting the original one makes the same angle with two parallel lines. Suppose the parallels are given to make the same angle with the line AB, and that CD is another line meeting them (Fig. 13). Let $\angle ACD = \beta$. Then

$$\angle EAC + \angle ECA = \pi - \angle AEC,$$

$$\angle EBD + \angle EDB = \pi - \angle BED.$$

But $\angle EAC = \angle EBD.$

Therefore $\angle ECA = \angle EDB.$

6·49. If three lines AB, AC, AD are all perpendicular to AO, they are in a plane. For if we make $OA' = OA$, we have easily $BO' = BO$, $CO' = CO$, $DO' = DO$. Hence B, C, D are all in the plane determined by O and O'.

6·50. Consider any two points L, M and a straight edge OP. Suppose points A on OP, B on OL, C on OM to have been found such that BA, CA are perpendicular to OP. Let

$$OL=r, \quad OM=r',$$

$$\angle LOP=\theta, \quad \angle MOP=\theta',$$

Then

$$LM^2=r^2+r'^2-2rr'\cos LOM, \quad (1)$$

and

$$\cos LOM = \cos\theta\cos\theta'$$

$$+\sin\theta\sin\theta'\cos BAC \quad (2)$$

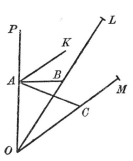

Fig. 14

by 6·40. If AK is any other straight edge through A perpendicular to OP, then K, A, B, C are in a plane, by 6·49. Let

$$\angle KAB=\phi, \quad \angle KAC=\phi'.$$

Then

$$\angle BAC=\phi'-\phi \quad (3)$$

and

$$LM^2=(r\cos\theta-r'\cos\theta')^2+(r\sin\theta\cos\phi-r'\sin\theta'\sin\phi')^2$$

$$+(r\sin\theta\sin\phi-r'\sin\theta'\sin\phi')^2 \quad (4)$$

If we now define $x=r\sin\theta\cos\phi$, $y=r\sin\theta\sin\phi$, $z=r\cos\theta$,

$$x'=r'\sin\theta'\cos\phi', \quad y'=r'\sin\theta'\sin\phi', \quad z'=r'\cos\theta' \quad (5)$$

we have

$$LM^2=(x-x')^2+(y-y')^2+(z-z')^2. \quad (6)$$

Thus distance has been expressed in the standard form appropriate to Cartesian co-ordinates.

The angles ϕ and ϕ' are independent of the position of A provided AK is always taken in the same plane. The angle between two planes is defined as the angle between two lines in them perpendicular to the common line, and is constant by the corollary to 6·40.

The above definition of Cartesian co-ordinates is applicable in all cases where it is possible to find the distances and bearings of the marks, whereas the usual definition is not applicable unless we can actually reach the projections on the three co-ordinate axes. We have still to show that our x, y, z are the same as the usual co-ordinates when these can be measured.

6·51. If we put $x = lr$, $y = mr$, $z = nr$, and consider two marks L, M given by $(x_1, y_1, z_1)(x_2, y_2, z_2)$, we have from 6·50 (2)

$$\cos LOM = l_1 l_2 + m_1 m_2 + n_1 n_2.$$

6·52. At any point of OP, $\theta = 0$, and therefore $l = 0$, $m = 0$, $n = 1$. If OQ is perpendicular to OP, it appears from 6·51 that $n = 0$ at Q. If also Q is in the same plane as OAK, $\phi = 0$ at Q, and $m = 0$. Hence at Q $l = 1$, $m = 0$, $n = 0$. If at R, $\theta = \frac{1}{2}\pi$, $\phi = \frac{1}{2}\pi$, we have at R, $l = 0$, $m = 1$, $n = 0$. Thus OQ, OP, OR are the co-ordinate axes as usually understood.

6·53. If $L(x, y, z)$ is another point, the angle between OL and OQ is given by
$$\cos LOQ = l \cdot 1 + m \cdot 0 + n \cdot 0,$$

and therefore by 6·39 the projection of OL on OQ is rl or x. Similar results hold for the projections on the other axes. This gives the identification required. If any of l, m, n are negative, the corresponding foot of the perpendicular from L on the axis lies on the production of the axis beyond O.

6·54. If now a plane is determined by two points (a, b, c), (a', b', c'), we have for all points on the plane

$$(x-a)^2 + (y-b)^2 + (z-c)^2 = (x-a')^2 + (y-b')^2 + (z-c')^2, \qquad (1)$$

that is,

$$2(a-a')x + 2(b-b')y + 2(c-c')z = a^2 + b^2 + c^2 - a'^2 - b'^2 - c'^2. \qquad (2)$$

Hence a plane has an equation of the first degree in the co-ordinates. Conversely if we are given an equation of the first degree, which in general involves three independent parameters, we can (in a triply infinite number of ways) assign the six co-ordinates of O and O' to make (2) fit it. Thus every equation of the first degree represents a plane.

It follows that a set of collinear points satisfy a pair of equations of the first degree. Also if a plane has the equation

$$Ax + By + Cz + D = 0 \qquad (3)$$

and $P(x_1, y_1, z_1)$, $Q(x_2, y_2, z_2)$ are two points satisfying (3), then

$$R\left(\frac{m_1 x_1 + m_2 x_2}{m_1 + m_2}, \quad \frac{m_1 y_1 + m_2 y_2}{m_1 + m_2}, \quad \frac{m_1 z_1 + m_2 z_2}{m_1 + m_2} \right)$$

also satisfies (3).

If then P and Q are common to two planes, the point R is also common to the two planes. If we call $R(x, y, z)$ we see that (x, y, z) satisfy

$$\frac{x - x_1}{x_2 - x_1} = \frac{y - y_1}{y_2 - y_1} = \frac{z - z_2}{z_2 - z_1}, \tag{4}$$

the usual form of the equations of a straight line.

From these results the usual analytic development can be carried out.

6·6. The foregoing theory has been developed from the notion and properties of distance alone. Most of the propositions inferred are verifiable by experiment, and have, of course, actually been verified. But other appliances exist that often enable us to supplement the theory and extend its practical application. The chief is the protractor or graduated circle, which we have already discussed in 6·0, and gives a direct measure of an angle.

In an observing tube the crossed wires at the eye, a distant pair in the instrument and the object mark A can be placed in a line by the test of coincidence of visual direction. Another distant pair Y is placed in line with O and a second object mark B. Then the angle AOB is the same as the angle XOY, which can be measured by a protractor. The sextant is based on a modification of this principle. The theodolite effectively consists of a telescope and two protractors, and measures two angles corresponding to the θ and ϕ of 6·50; θ is measured from the upward vertical and ϕ from a plane containing the upward vertical and the north.

In general three measured data are needed to specify a position. These may be any functions of (r, θ, ϕ) or of (x, y, z) provided that none of them is a function of the other two. But our principle of simplicity gives reason for regarding the Cartesian co-ordinates as the physically fundamental ones. The directly recognized entities are straight lines and distances, and a plane is a notion that arises directly out of distance. Now Cartesian co-ordinates have the following simple features possessed by no others. Any plane is expressed by an equation linear in the co-ordinates. Any straight line is expressed by a pair of linear equations; and the co-ordinates of any point on the line are weighted means of those of two given points on the line. The square of the distance between two points is the sum of the squares of the differences between the co-ordinates.

No other general relations of comparable simplicity hold for any other type of co-ordinates.

Another feature of Cartesian co-ordinates may be regarded as the beginning of the notion of relativity. The origin can be taken at any convenient point; and given the origin there is still a three-fold arbitrariness in the directions of the axes. Any such change of the reference system changes the values of the co-ordinates of a given mark; but it does not change the distance between two marks. A change of origin without change of the directions of the axes does not alter the differences of the co-ordinates; a change of direction of the axes applies linear transformations to the co-ordinates. We have thus considerable latitude in the choice of reference system even when we restrict ourselves to Cartesian co-ordinates. All such systems are equivalent in the sense that the equations take the same forms in all of them, though the actual values of the quantities entering may be different.

6·7. The chief experimental law is that adopted in 6·22. This can be verified in a laboratory with metre scales, and with due care the measures can be made to about 0·1 mm., say to 1 part in 10^4 for the greatest lengths available and 1 in 10^3 for intermediate ones. What we have done is to treat it as exact and develop its consequences. This has involved extrapolation to both larger and smaller quantities.

At the best the errors in the data are those due to the step of the scale, and can be reduced by taking many observations, but they can never be made zero. We can, however, set marks that would, according to our rules, form an isosceles right-angled triangle, and this implies the existence of lengths whose ratios are irrational. The hypothesis that the whole range of real numbers is available for ratios frees this from any mathematical difficulty. But do we need all this range for physical purposes? We might, for instance, say that only rationals can arise. Since the uncertainty is never zero, and there are rationals arbitrarily near any real number, observation can never disprove the statement that all ratios of lengths are rational. In that case the isosceles right-angled triangle might not exist. Alternatively, the law §6·22(5) might be false for irrational ratios, and observation cannot disprove this either.

If we were restricted to rational ratios, the natural interpretation would be that all lengths are exact multiples of some least possible length. It is known that

$$(m^2 - n^2)^2 + (2mn)^2 = (m^2 + n^2)^2,$$

and if the sides of a triangle are $m^2 - n^2$, $2mn$, $m^2 + n^2$ times a standard length, where m and n are integers, Pythagoras's theorem holds and the triangle is right-angled. But $m^2 - n^2$ cannot be equal to $2mn$ for any integers m, n. Conversely for all right-angled triangles whose sides are integral multiples of a standard, m and n exist. We can, however, take m and n to make the ratio of these expressions as near 1 as we like. We could therefore have a triangle that is indistinguishable by measurement from an isosceles right-angled triangle.

However, a principle of this sort would involve the minimum length explicitly, and therefore the laws would have to contain it explicitly. But a general principle of this sort is useless to us. We can make no calculations that depend on a fundamental length unless we know what that length is. We could estimate it only by way of discrepancies between the law and still more accurate measurements than we can make now, and if we do so we shall be able to take it into account. To introduce it at the present stage would be an introduction of a new parameter with no means whatever of estimating it or testing its significance, and consequently would not be a practical method of progress. The possibility should not be forgotten, but to insist on it now would obstruct progress and not encourage it. Thus the use of irrational ratios of lengths does not definitely assume that such ratios really occur in physics; it only provides a background which saves a great deal of calculation that would be useless at present (that is, finding rational solutions of equations where the existence of real solutions presents no difficulty.)

We refrained from making the assumption that a line can have any length, however great. But in 6·45 the quotient

$$O'P \cos O'PQ - OP \cos OPQ$$

might, as far as we have said, be as small as we like, and thus PR might be arbitrarily large. This does not contradict our remark that there is always a practical limit to the length of a straight edge; it is consistent to say that we can calculate PR without actually being able to reach R and set a physical mark there. Similar considerations apply to the discussion of parallels. We have given a practical test in terms of measurable angles (which, in fact, is very like Euclid's). Non-intersection is not taken as a definition of parallelism; what is proved is that lines that are parallel according to the definition cannot intersect even if they are allowed to have arbitrarily great lengths.

It is formally possible that the laws might cease to be accurate at very large distances. This would be shown, for instance, if experiments that determine the position of a point R gave results that differ significantly, regard being had to the uncertainties. In that case we should need to modify the laws. But, as for small distances, we cannot usefully do so until we have evidence that would tell us how and how much we should modify them. Here we approach problems relating to the size of the Universe.

The conclusion is the same in both cases. We should maintain the simple law and extrapolate it by calculation whenever we can. That is both the best way of finding useful results and the best way of finding out departures from the law if it is false. In other words, we should never modify a simple law without positive reason.

Now this is precisely what we inferred from the theory of probability in Chapter II. Any law must contain a statement of probabilities of errors of various amounts, and laws must be tested in an order, those with the least number of adjustable parameters being taken earlier. The theory of probability also puts the rule for testing the significance of departures into a quantitative form; it thus provides the criterion that is needed to say what we mean by 'positive reason' in the last paragraph.

On the question of the verified accuracy of the laws of mensuration, I consulted Mr J. E. Jackson, who replied as follows:

'There are several good examples of survey base-lines connected up by triangulation.

'In the recently completed retriangulation of Great Britain there are geodetic bases at Ridgeway and Caithness, the former 11·3 km. long, the latter 24·8 km. They are exceptionally far apart, nearly 500 miles, and a chain of about 50 triangles would have to be computed to join them up. Calculating from the Ridgeway value, the Caithness base comes out 12 cm. (1 in 200,000) longer than the directly measured value.

'The Sudan Survey reports on the triangulation along the Nile Valley quote base to base misclosures of about 1 in 100,000.

This seems to be of the order of things in modern geodetic surveying.

'In Ceylon, our two modern bases were each about 9 km. long, separated by a distance of about 130 miles or 16 triangles. The misclosure was 1 in 110,000, with a rather inferior quality of triangulation.

'The modern theodolite is better than 1″, I think. Triangle misclosures in first-class work nowadays average less than 1″. The Sudan report referred to above gives 0·2″ as the "probable error of an adjusted direction", which presumably means about 0·3″ for an adjusted triangle.'

Thus laws derived for distances of the order of metres are extrapolated to 10^5 or more times the original scale, and still verified to one part in 10^5 or more.

CHAPTER VII

NEWTONIAN DYNAMICS

Nature, and Nature's laws, lay hid in night;
God said, Let Newton be! and all was light.

POPE

7·0. Many rigid systems, in the sense of the previous chapter, exist. The criterion for a rigid system is that the distances between recognizable marks in it do not change with the time. If one distance in a system and all angles, as tested by optical instruments with graduated circles, do not change with the time, we still call the system a rigid system. Over considerable intervals of time most of the objects in this room constitute a rigid system. The angular distances between stars, as observed from the Earth, vary with the time so little that years are required to detect alteration even with the best measuring instruments. If then we consider a system of lines through a given point, and each directed toward a star, that constitutes a rigid system.

When distances change with the time we are in a new realm, called dynamics. The marks in one rigid system may change their distances and directions from those in another rigid system. Thus a theodolite and the objects on the Earth within its field of view constitute a rigid system; the stars and an equatorial telescope with the clockwork going constitute another; but the directions of stars change with respect to the theodolite, and those of objects on the Earth change with respect to the equatorial.

Objects whose distances and directions with respect to a rigid system vary with the time are said to have motion relative to the system. Distance and angle have so far been considered only when they are constant for a given set of two or three marks. But even when they vary they can still be measured. We can specify the position of a particle sliding down a curve by the mark on the curve that the particle is passing over. We can specify the direction of a planet by pointing an equatorial telescope toward it, and reading the right ascension and declination on the graduated circles; or we can photograph the region of the sky where it is and measure its angular distance from neighbouring fixed stars, just as we can measure the angular distances between these stars themselves. In such a case as the ascent of a pilot balloon, observed with two theodolites, we can actually observe the directions from two positions simultaneously and determine the position of the balloon at each instant of observation just as for a fixed object. In dynamics we are therefore dealing with cases where distances, and those entities that we have found to be related to them, are still measurable, but are now functions of the time instead of being constant. We have also to consider cases where bodies are deformable; distances between marks on the same body may change, or marks formerly collinear may cease to be so. This phenomenon belongs to the branch of mechanics called elasticity. Much of the work on this subject considers a system in different states; it may in suitable circumstances be able to exist in either state for long intervals of time, but change of the conditions may lead distances within it to change from one set of values to another. We then speak of the conditions in either state as static.

The elementary subject of statics is really composed of parts of several subjects. In the common balance the pans, weights, the beam, separately behave as rigid bodies; but as the beam tilts the distances between points on, say, one pan and the opposite end of the beam change. The system as a whole is not rigid, and its theory depends on a study of the interrelations between the changes of distances and the bodies put on the pans. The theory, that is, concerns relations between different rigid bodies. The 'light inextensible string' is an idealized approximation to an actual string. An actual string opposes little resistance to bending and a great deal to changes of length; the idealized string opposes none at all to bending and is completely inextensible. The elastic string resembles the light inextensible string in offering little resistance to bending, but it can be stretched considerably under moderate force. Statics is the study of bodies or systems of bodies when there is no relative motion of their parts, and such states are classified together as equilibrium.

Dynamics on the other hand deals with bodies in relative motion. It has in common with statics the laws of mensuration and the new notion of *force*. In the first place force is recognizable by bodily sensation. With practice some people can even estimate it fairly well quantitatively, like those at village shows who can estimate the weight of a basket of vegetables to about 5 per cent. We can recognize force when we stretch a string, lift a weight, or throw a ball. Some accounts of dynamics try to dispense altogether with this primitive sensation, chiefly, I think, because it is not accurately measurable. This seems to me to go too far; force resembles distance in being roughly estimable without instruments, but it can be much more accurately measured with them. To reject the sensation of force would be a denial of the principle that science must be ready to admit and examine *any* information whatever; that is just what distinguishes science from propaganda.

One important feature of the sensation of force is that it occurs in both statical and moving systems, and thus provides a way of connecting the two. In statics force can be measured by the stretching of a standard spring; in dynamics it is measured as the product of mass and acceleration; and the sensation of force is the most direct reason we have for supposing that the two quantities measured are really the same thing.

There is a difference of opinion among experts as to whether it is better to begin with statics or dynamics. For elementary teaching statics is preferable because many of its problems can be solved without the calculus; we are, however, supposing that the principal methods of the calculus are known already. Against the definition of force by way of elasticity is the fact that the proportionality of force to extension is usually inaccurate at large extensions, and consequently the range of force that can be treated is comparatively small. Another objection to beginning with statics is that its problems fall into two classes, according as the position of equilibrium is unique or not. If it is unique, it usually arises as the result of the dying away of an oscillation, in which case the statical problem is only a special case of the corresponding dynamical one. Often, especially in problems involving solid friction, it is not unique, and the estimation of the force is impossible without some further assumption. For solids in contact it is necessary to assume a reaction, which there is no direct means of measuring; and the reaction is under suspicion of having been introduced to make the laws true.

The disadvantage of dynamics is that the most elementary motion, that of a body falling under gravity, is too fast to be accurately observed in ordinary laboratory conditions except by special devices such as the pendulum and Atwood's machine for reducing the accelerations; and these have the same disadvantage as the statical method, that they introduce non-measurable reactions.

7·1. In any case both types of problem must be studied, and elasticity and dynamics combine in problems of vibrations of strings and elastic waves. The great advantage of taking dynamics first is that we can proceed directly to dynamical astronomy, which is not only the oldest but the most accurate of all quantitative sciences after mensuration.

The first point to notice is that material connexion between bodies is antagonistic to relative motion. Even if there is relative motion to begin with, as in the case of a body projected along the floor, it soon stops. In order, therefore, to deal with one thing at a time as far as possible it is best to consider first bodies that have the slightest possible material connexion with the frame of reference.

This condition is satisfied fairly well by bodies in mid-air, and very well by the heavenly bodies. In the first place we take axes fixed in the Earth, and find that the motion, at any rate for massive bodies, is well represented by the differential equations

$$\ddot{x} = 0, \quad \ddot{y} = 0, \quad \ddot{z} = -g, \tag{1}$$

where x and y are measured horizontally and z vertically upward. The derived magnitude g is nearly constant.† Now by simultaneous observations of the stars from different points of the Earth's surface we find that the direction of the vertical (judged by the plumb-line or a trough of liquid) is not the same everywhere with reference to the directions of the stars, but points nearly to a fixed point in the Earth, which we call the centre. If we take new Cartesian co-ordinates with respect to the centre of the Earth as origin, the equations (1) lead to

$$\ddot{x} = -g\frac{x}{r}, \quad \ddot{y} = -g\frac{y}{r}, \quad \ddot{z} = -g\frac{z}{r}, \tag{2}$$

where r is measured from the centre. This is a more general form than (1). We call the first time-derivatives of the Cartesian co-ordinates the components of relative velocity, and the second derivatives the components of relative acceleration. There is a strong suggestion that the accelerations are functions of the co-ordinates alone, not of the velocities.

Now consider the motions of the two stars that form a double

† In recent work by C. Volet and A. H. Cook a freely falling body has been used to give a direct determination of gravity; the accuracy, with modern methods of timing, seems better than that of the best work with pendulums.

star. We take an axis of x in a direction fixed with regard to the directions of the majority of the stars, and such that the double star is always near it. The axes of y and z are taken in two directions perpendicular to each other and to that of x. We observe the angles between the directions of each separate star and the two planes of xy and xz; or, what is equivalent, we take the point of intersection, with a plane plate perpendicular to the x axis, of the line joining the centre of the object glass to the star. It is found that as time goes on the points given by the two stars describe similar ellipses. If we put $y/x=p$, $z/x=q$, and use suffixes 1 and 2 for the two stars, the variations of p and q in the same interval of time are always opposite in direction and in a constant ratio, except for a uniform velocity shared by both stars. If we proceed to second derivatives to remove the uniform part of the rate of change, we have

$$\frac{\ddot{p}_1}{\ddot{q}_1}=\frac{\ddot{p}_2}{\ddot{q}_2}=\frac{p_2-p_1}{q_2-q_1}. \tag{3}$$

Now p and q are always small. If, as is reasonable, we suppose that the motion in the line of sight is comparable with that across the line of sight, we can infer

$$\frac{\ddot{y}_1}{\ddot{z}_1}=\frac{\ddot{y}_2}{\ddot{z}_2}=\frac{y_2-y_1}{z_2-z_1} \tag{4}$$

and if we further assume that an observer looking at the star from a different direction would find similar results we can extend (4) to the x component and generalize the equations in the form

$$\frac{\ddot{x}_1}{x_2-x_1}=\frac{\ddot{y}_1}{y_2-y_1}=\frac{\ddot{z}_2}{z_2-z_1};\quad \frac{\ddot{x}_2}{x_2-x_1}=\frac{\ddot{y}_2}{y_2-y_1}=\frac{\ddot{z}_2}{z_2-z_1}. \tag{5}$$

The components of acceleration are in the ratios of the differences of the co-ordinates; in other words, the accelerations of the bodies are in the line joining them. Further, we can choose a ratio of two positive quantities μ_1, μ_2 such that

$$\mu_1\ddot{x}_1+\mu_2\ddot{x}_2=0,\quad \mu_1\ddot{y}_1+\mu_2\ddot{y}_2=0,\quad \mu_1\ddot{z}_1+\mu_2\ddot{z}_2=0. \tag{6}$$

The relative accelerations are not noticeable for distant bodies. Hence we can say that the acceleration of each star is *due to* the proximity of the other star.

Similar results are found for most of the satellites of the planets; they are consistent with the acceleration in each case being directed toward the centre of the planet.

Now consider the acceleration of the Moon. To a first approximation the Moon describes a circle relative to the Earth with radius a and angular velocity n, where

$$a = 3 \cdot 8 \times 10^{10} \text{ cm.}, \quad n = 2\pi/27 \cdot 3 \text{ days.}$$

The acceleration is an^2 towards the centre of the Earth, and is $0 \cdot 273$ cm./sec.^{2}. Now a falling particle at the Earth's surface has acceleration 980 cm/sec.^{2}, which is nearly 3600 times the acceleration of the Moon. The distances are in the ratio $1 : 60$ nearly, so that the accelerations are nearly inversely as the squares of the distances.

7·2. These few facts relating to freely moving bodies suggest the following summary:

A body has an acceleration towards a neighbouring body, proportional in magnitude to the inverse square of the distance.

The accelerations that two bodies produce on each other are in a ratio independent of the time and of the positions.

The first of these laws can be written in the form

$$\ddot{x}_1 = -\frac{\mu_2}{r^2}\frac{x_1 - x_2}{r}; \quad \ddot{y}_1 = -\frac{\mu_2}{r^2}\frac{y_1 - y_2}{r}; \quad \ddot{z}_1 = -\frac{\mu_2}{r^2}\frac{z_1 - z_2}{r}, \quad (1)$$

and the second then in the form

$$\ddot{x}_2 = \frac{\mu_1}{r^2}\frac{x_1 - x_2}{r}; \quad \ddot{y}_1 = \frac{\mu_1}{r^2}\frac{y_1 - y_2}{r}; \quad \ddot{z}_2 = \frac{\mu_1}{r^2}\frac{z_1 - z_2}{r}. \quad (2)$$

This family of differential equations can be solved exactly. They are found to imply the following consequences:

A point with co-ordinates $\bar{x}, \bar{y}, \bar{z}$ given by

$$(\mu_1 + \mu_2)(\bar{x}, \bar{y}, \bar{z}) = (\mu_1 x_1 + \mu_2 x_2, \mu_1 y_1 + \mu_2 y_2, \mu_1 z_1 + \mu_2 z_2) \quad (3)$$

moves with uniform velocity in a straight line. We call this point the centre of mass (or centroid) of the two particles.

Relative to this point the bodies describe ellipses, the ellipses being similar but having their axes in opposite directions, and the centroid being in a focus of each.

The line joining the centroid to either body sweeps out in any interval of time an area proportional to that interval.

The mean distance a between the bodies being defined as the mean of their greatest and least distances apart, and the mean motion n as 2π divided by the time of describing the orbit,

$$n^2 a^3 = \mu_1 + \mu_2. \tag{4}$$

These results express Newton's solution of the Problem of Two Bodies. It is found to describe accurately the motions of double stars. Only motions at right angles to the line of sight being measurable,† what we actually verify is that the observed movements agree with the projections of movements that follow the laws. In other words, the behaviour of two of the three variables determined by the solution is completely verified; this is regarded as strong confirmation of the laws, which are then extended to the third variable.

When we consider the motions of satellites about their primaries, a similar solution is found to fit the relative motion, as to the two measurable co-ordinates. Also the motion of the planet itself is sensibly uniform over intervals amounting to several periods of revolution of the satellites. It appears therefore that the centre of mass of the planet and any satellite is nearly coincident with the centre of the planet, and therefore if μ_1 refers to the planet and μ_2 to the satellite, μ_2/μ_1 is always small and $\mu_1 + \mu_2$ is practically μ_1. But for different satellites of the same planet we get a further check; the quantity $n^2 a^3$ is nearly the same for all. This shows that μ_1 is a property of the planet.

Coming now to the motions of the Sun and planets, we can observe in each case only directions as seen from the Earth with reference to the stars, except in the case of the Sun, where we can estimate the variation of its distance by measuring its angular diameter from time to time. In this case, then, we can check all three co-ordinates, and we find definitely that the motion agrees satisfactorily with elliptic motion relative to the Earth as a focus; and therefore equally well with elliptic motion of the Earth relative to the Sun as a focus. For the other planets' motions relative to the Sun the previous rules give nearly the correct directions. The values of $n^2 a^3$ for the planets, including the Earth, are in close agreement, and this quantity therefore represents a property of the Sun rather than the Earth.

---

† That is to say, in terms of the considerations of direction that we have used so far. Velocities in the line of sight can be measured by means of the Doppler effect, and agree with the laws, but we are not yet in a position to discuss the theory of that effect.

We have presented the argument as it might be developed by a modern geodesist, armed with a good telescope, but unaware of previous work on the subject. He would discover double stars, and find their treatment fairly easy; he would observe the planets, but in the first place would find their motions baffling, on account of the large disturbances of direction due to the Earth's motion. Until this was disentangled little progress would be possible. Nevertheless, the discovery was actually made from the planets. Jupiter's greater satellites were discovered in 1609, the year of the publication of Kepler's first two laws, and the study of double stars is fairly modern. With naked-eye observations the variation of the angular diameter of the Sun would be imperceptible.

The historical accounts of the discovery of the laws seem to me to be obscure on several crucial points. I have not read the works of Ptolemy, Copernicus and Kepler in the original, and see no prospect of doing so, but it would be helpful if historians of science would explain some points that seem to me to be obscure. In the first place, Ptolemy is said to have produced a geocentric system of planetary motions, involving 80 epicycles; and Copernicus to have produced a heliocentric system, which involved only 34 epicycles. The system of Copernicus is said to have been adopted because it gave better agreement with observation.† Kepler, again, is said to have improved on Copernicus by introducing elliptic motion instead of epicycles. What does not appear to be remarked is that for small eccentricities elliptic motion can be well represented by epicycles. If

$$r^2\dot\theta = a^2 n, \quad r = a(1 + e\cos nt),$$

we find (with $\theta = 0$ at $t = 0$), neglecting e^2,

$$\theta = nt - 2e\sin nt$$

and

$$x = r\cos\theta = a(\cos nt + \tfrac{3}{2}e - \tfrac{1}{2}e\cos 2nt),$$

$$y = r\sin\theta = a(\sin nt - \tfrac{1}{2}e\sin 2nt).$$

To the first order in e, therefore, the motion can be regarded as composed of (1) a circular motion in a circle of radius a with angular velocity n, (2) a displacement of the centre by $\tfrac{3}{2}ea$, (3) a motion in a circle of radius $\tfrac{1}{2}ea$ with angular velocity $2n$.

† H. Butterfield, *The Origins of Modern Science* (1950), p. 26.

O. Gingerich† writes, 'In its classical form the Ptolemaic mechanism for Venus, Mars, Jupiter and Saturn consisted of a single epicycle moving on a larger circle with uniform speed about an off-centre point called the equant. Had the parameters been well determined the theory could have predicted planetary longitudes within a few minutes of arc, except for Mars, which at times would have had an error up to half a degree. Although Copernicus eliminated the large epicycles by going to a heliocentric model, he invoked a small epicycle for each of these planets, ending up with a less flexible and less accurate mechanism, but with more circles than Ptolemy!'

The above passage seems to depend on the fact that the above formulae lead to

$$\frac{y}{x - 2ae} = \tan nt$$

with no term in e; that is, the direction of the planet from the unoccupied focus revolves at a constant rate, to order e. This motion is not, of course, in a circle, but pre-Copernican astronomers were concerned only with direction. But in any analysis the epicycle for the Sun should appear in all those for the planets, and it is hard to see how the repetition of the annual period could have been missed.

Copernicus was working with six planets. Thus if eccentricities were neglected the best fit would be by six circular motions; if the first power of the eccentricity is retained, six displacements of the centres and six epicycles would be determinable. In Ptolemy's representation the motion of the Sun relative to the Earth would be added to that of every planet, and would appear in the first approximation. It is larger than the effect of any eccentricity. So apparently if Ptolemy had made the best approximation to the observed motions by using two circles for each planet, he would have found that one circle of the same radius and angular velocity appeared for every planet. If he did not get this, why not? And if he did, why does nobody mention it, since it is plainly the most direct evidence that a motion of the observer was being interpreted as motion of the planets? Had he used one more circle and displacements of the centre for each planet he should also have found the eccentricity

† *American Scientist*, **55**, 1967, 88–95.

terms for the Sun repeating themselves. With six terms for each planet (not counting the Earth) and three for the Sun, 33 in all, he should have been able to represent the whole motion to the first order in the eccentricities.

There is another mystery about the discovery of the inverse square law. It was often mentioned in correspondence before Newton's publication, but Newton's only statement about the original suggestion appears to have been in a letter, in which he said that it was based on an analogy with the intensity of light. However, Kepler's laws were known. If we take only the first law and the third, but in a somewhat restricted form, we can say: the motion of a planet, whatever its initial velocity, is in a plane that passes through the Sun; and n^2a^3 would be the same for planets at any distance if they move in circular orbits. From the first statement it follows, since the only line common to all planes through the planet and the Sun is the line joining them, that the acceleration must be along that line. Now for a circular orbit, if we write

$$x = a \cos nt, \quad y = a \sin nt,$$
$$\ddot{x} = -n^2x, \quad \ddot{y} = -n^2y,$$

and therefore the acceleration is radial and of magnitude n^2a. But from the third law n^2a^3 is the same for all a; hence the acceleration is proportional to a^{-2}.

This argument was well within the powers, not only of Newton, but of Leibniz, James Gregory, James and John Bernoulli, Halley, and even Hooke. It seems reasonable to suppose that it was in fact well known and somehow escaped publication until Newton achieved the much more difficult task of showing that the inverse square law implies motion in a conic.†

7·3. We have seen that a body in the neighbourhood of a second body has an acceleration towards the second body, inversely proportional to the square of the distance, the constant factor being a property of the second body alone. What happens when there are several bodies in the neighbourhood? We really have the answer

---

† I consulted Professor H. W. Turnbull on the point; he says that the inverse square law for planetary orbits can be traced back to Bullialdus, *Astronomica Philolaica* (Paris, 1645), p. 23; and Newton refers to Bullialdus in a letter to Halley of 1686 June 20.

already in the motions of satellites; for while a satellite is moving about a planet it is also sharing in the motion of the planet about the Sun. If the acceleration was merely toward the primary, while the primary is moving about the Sun, the primary would leave the satellite behind.† The satellite must also have an acceleration toward the Sun, which is nearly the same as that of the primary because they are at nearly the same distance from the Sun. We must therefore generalize the law to the case where n bodies are moving in one another's neighbourhood. We say that any one body has an acceleration toward each of the others, whose components are given by our law; and the total component acceleration in any direction is the sum of the components in that direction given by the other bodies separately. Formally, if we consider the αth body,

$$\ddot{x}_\alpha = \Sigma' \frac{\mu_\alpha}{r_{\alpha\beta}^3}(x_\beta - x_\alpha), \text{ etc.,} \tag{1}$$

where the suffix β refers to another body of the system, $r_{\alpha\beta}$ is the distance between the bodies specified by α and β, and Σ' denotes summation for all values of β except α.

We notice that

$$r_{\alpha\beta}^2 = (x_\alpha - x_\beta)^2 + (y_\alpha - y_\beta)^2 + (z_\alpha - z_\beta)^2 \tag{2}$$

and
$$\frac{x_\alpha - x_\beta}{r_{\alpha\beta}^3} = -\frac{\partial}{\partial x_\alpha}\frac{1}{r_{\alpha\beta}} = \frac{\partial}{\partial x_\beta}\frac{1}{r_{\alpha\beta}}, \tag{3}$$

with similar relations. Then

$$\ddot{x}_\alpha = \frac{\partial}{\partial x_\alpha}\Sigma'\frac{\mu_\beta}{r_{\alpha\beta}}. \tag{4}$$

We can now multiply (4) by μ_α and add for all values of α. Then the pair of particles specified by α and β make a contribution

$$\left(\frac{\partial}{\partial x_\alpha} + \frac{\partial}{\partial x_\beta}\right)\frac{\mu_\alpha\mu_\beta}{r_{\alpha\beta}} = 0, \tag{5}$$

and therefore
$$(\Sigma\mu_\alpha)\ddot{x} = \Sigma_\alpha \mu_\alpha \ddot{x}_\alpha = 0. \tag{6}$$

Hence the centre of mass of the system moves with uniform velocity in a straight line.

† This is serious; the acceleration of the Moon toward the Sun, for instance, is about twice its acceleration toward the Earth.

If we multiply (4) by $\mu_\alpha \dot{x}_\alpha$, $\mu_\alpha \dot{y}_\alpha$, $\mu_\alpha \dot{z}_\alpha$ and add, and then add for all particles, the left sides give

$$\frac{d}{dt} \Sigma \tfrac{1}{2} \mu_\alpha (\dot{x}_\alpha^2 + \dot{y}_\alpha^2 + \dot{z}_\alpha^2). \tag{7}$$

The contribution to the right from the terms containing $r_{\alpha\beta}$ is

$$\mu_\alpha \mu_\beta \left(\dot{x}_\alpha \frac{\partial}{\partial x_\alpha} + \dots + \dot{x}_\beta \frac{\partial}{\partial x_\beta} + \dots \right) \frac{1}{r_{\alpha\beta}} = \mu_\alpha \mu_\beta \frac{d}{dt} \frac{1}{r_{\alpha\beta}} \tag{8}$$

since $r_{\alpha\beta}$ is a function of the two sets of co-ordinates only.

Hence we have

$$\left[\Sigma \tfrac{1}{2} \mu_\alpha (\dot{x}_\alpha^2 + \dot{y}_\alpha^2 + \dot{z}_\alpha^2) \right]_{t_0}^{t_1} = \left[\Sigma\Sigma \frac{\mu_\alpha \mu_\beta}{r} \right]_{t_0}^{t_1}, \tag{9}$$

each pair being counted once in the sum on the right.

This is a very remarkable result. For the left side is simply the change of $\Sigma \tfrac{1}{2} \mu_\alpha v_\alpha^2$, where v_α is the resultant velocity of the αth particle. The quantities μ_α are properties of the various bodies and independent of their position. Thus (9) expresses a relation between the elementary notions of distance and time alone, and is independent of the particular set of axes used. Further, it remains true if the origin is taken to have any constant velocity, so long as the axes remain parallel to their original direction. For if we put

$$x_\alpha = x_\alpha' + ut, \quad \bar{x} = \bar{x}' + ut, \tag{10}$$

$$\frac{d}{dt} \Sigma \tfrac{1}{2} \mu_\alpha \dot{x}_\alpha^2 = \frac{d}{dt} \Sigma \tfrac{1}{2} \mu_\alpha (\dot{x}_\alpha'^2 + 2\dot{x}_\alpha' u + u^2). \tag{11}$$

But $\quad \dfrac{d}{dt} \Sigma \mu_\alpha \dot{x}_\alpha = \dfrac{d}{dt} (\Sigma \mu_\alpha) \dot{\bar{x}} = 0, \quad \dfrac{d}{dt} \Sigma \mu_\alpha u^2 = 0, \tag{12}$

and the expression is simply $\dfrac{d}{dt} \Sigma \tfrac{1}{2} \mu_\alpha \dot{x}_\alpha'^2$. In particular we can take $u = \dot{\bar{x}}$, thus referring the whole motion to the centre of mass of the system. The principle is a case of the conservation of energy.

We may notice also that in (1) the differences between the accelerations of the particles are functions of the distances alone, and are therefore the same if the origin has a velocity, or even an acceleration. This is Newton's form of the principle of relativity, and it is the basis of the later attempts to state the equations in a form that contains no reference to quantities other than those that can be actually measured. We may remark at once, however, that if there are n bodies, only $n-1$ differences of position can be assigned independently, and practical ways of making use of this simplification do so by abandoning the formal similarity of the equations for the different particles.

The equations (4) are equivalent to

$$\mu_\alpha \ddot{x}_\alpha = \frac{\partial}{\partial x_\alpha} \Sigma\Sigma' \frac{\mu_\alpha \mu_\beta}{r_{\alpha\beta}}, \text{ etc.} \tag{13}$$

The generalization (4) makes a great improvement in our representation of the motions within the solar system. Kepler's laws give a good first (or rather second) approximation to the motions of the planets; their application to the motions of the satellites relative to the planets also gives a good approximation. But there are outstanding discrepancies. There are periodic inequalities in the Moon's longitude with amplitudes of the order of a degree; others in the longitudes of the planets of the order, in extreme cases, of considerable fractions of a degree; there is a long-period disturbance of Jupiter and Saturn with a period of nearly 900 years and an amplitude of nearly a degree. In addition the elements of the orbits show slow progressive changes, the major axes revolving with respect to the stars. For some satellites there are even more striking peculiarities. The result of allowing for the mutual influence of every pair of bodies in the system is that nearly all these inequalities are accounted for. Without further modification we can account closely for the motion of every major plant from Venus to Pluto, all the asteroids, and most of the satellites. The exceptions are for the most remote satellites of Jupiter and Saturn, for which the calculations are still incomplete.

The outstanding discrepancies all concern cases where the body whose motion we are considering is very near its primary. This fact suggests an explanation; for we have seen that the acceleration of a body is always towards the body that produces it. If the latter is remote, the lines joining the first body to all parts of the second are nearly in the same direction. But if the bodies are fairly close together these lines are not in the same direction. If the second body is a sphere its field will be symmetrical, but if not, the acceleration of the first body will not necessarily be directed toward a fixed point of the second. Such considerations do, as a matter of fact, account for most of the outstanding inequalities of the satellites. The only remaining inequality of importance concerns Mercury; its discussion is reserved till later.

When the disturbances due to other planets are taken into account the directions of the planets, as seen from the Earth, are predictable to about $1''$ (this will probably be improved when calculations in progress are completed). The disturbance of direction due to the motion of the Earth amounts to about $40°$ for Mars; the ratio of the distances is accordingly verified to about 7 parts in 10^6 and Kepler's third law to about 2 parts in 10^5. Similar accuracy is obtained for Venus. (The mean angular velocities are known to much higher accuracy.)

We have seen that the acceleration of any body can be considered as made up of contributions from other bodies, each of which can be said to be due to another particular body in the sense that it would be zero if the other body was not present. If then we denote the part of the acceleration of the αth particle due to the βth by $\ddot{x}_{\alpha\beta}, \ddot{y}_{\alpha\beta}, \ddot{z}_{\alpha\beta}$, we have

$$\mu_\alpha \ddot{x}_{\alpha\beta} + \mu_\beta \ddot{x}_{\beta\alpha} = 0, \text{ etc.} \qquad (14)$$

We can call the terms in this equation the respective forces of the bodies on each other, and we arrive at a result equivalent to Newton's third law, that action and reaction are equal and opposite.

The equation (14) is probably most directly verified in the solar system by the mutual perturbations of Jupiter and Saturn. The difficult thing about direct verification is that it requires independent estimates of the masses to exist, and the mutual perturbations of the two bodies considered to be large enough to be measured. In the case of Jupiter and Saturn the masses are known to about 1 part in 1000 from the motions of the satellites, and the mutual perturbations to higher accuracy. However, (14) is only a consequence of the general equations (13) and what really concerns us in astronomy is their verification as a whole. For various reasons, however, it is worth while to break the solution up into parts. The motion of each body can be represented by a series of terms whose periods are simple combinations of the periods of revolution, and the method most used is to analyse the observed motion into terms of the corresponding periods. The coefficients of these terms depend on the masses, and provide the best way of estimating these. The main point is that the estimates of the mass of Jupiter from the satellites, and from perturbations of Saturn and of various asteroids, are consistent and provide a verification that μ for Jupiter, as well as for the Sun, is a property of the attracting body alone.

We may remark that there is nothing conventional about the quantities μ. They are perfectly definite derived magnitudes. Thus (very nearly) μ for the Sun $= 4\pi^2 \dfrac{(\text{Earth's mean distance})^3}{(\text{1 year})^2}$. The relation to the laboratory definition of mass has not yet appeared.

The actual procedure in gravitational astronomy is to convert the equations (1) into equations for the differences of the accelerations of the other bodies and the Sun; thus the unknown acceleration of the centre of mass of the whole system disappears. The procedure up to (1) is of the nature of a preliminary search. What is really verified is the full set of equations for the differences of the co-ordinates.

7·4. One point that needs special mention is that the axes in 7·3 (1) are taken to be non-rotating. Thus we have had, for motion in a circular orbit, a solution

$$x = a\cos nt, \quad y = a\sin nt, \quad z = 0, \tag{1}$$

satisfying $$(\ddot{x}, \ddot{y}, \ddot{z}) = -\frac{\mu}{r^3}(x, y, z) \tag{2}$$

with $\mu = n^2 a^3$. But if we take axes of (x', y', z') rotating about the axis of z with angular velocity n, the co-ordinates in the two systems are connected by relations

$$x' = x\cos nt + y\sin nt; \quad y' = -x\sin nt + y\cos nt; \quad z' = z. \tag{3}$$

Then $x' = a$, $y' = 0$, $z' = 0$. But in this form the equations of motion in the form 7·3 (1) are not satisfied, for

$$\ddot{x}' = 0, \quad -\frac{\mu}{r^3}x' = -n^2 a. \tag{4}$$

The form 7·3 (1) cannot therefore be true for all sets of axes in relative rotation. If it is true for axes fixed in relation to the stars, it cannot be true for axes fixed in relation to the Earth. It is motion relative to the stars that we have been considering; for motion relative to the Earth a modification is needed, which is given in books on dynamics. It is actually found to be small for a projectile moving near the Earth's surface, really because the time of flight is so small that the Earth rotates by only a small angle during it. But the correction is appreciable in long-range gunfire and in a very long pendulum.

7·5. When we come to deal with bodies that are not moving freely we find a difference at once. If two bodies are in contact the relative acceleration disappears; the law of attraction undergoes serious modification at this stage. What are we to say about a book lying on the table? Two courses are available. The book is not in contact with the Earth; we may say that due to the Earth, the book has, as usual, the acceleration g downward, but owing to the proximity of the table it has an additional acceleration g upward, and the two cancel. Otherwise we may just say that the acceleration is zero and leave it at that. The second procedure is less trouble. It might even be argued that the reaction from the table is simply an invention to make the laws of motion true and has no physical meaning. Nevertheless the first procedure is adopted. It has the important feature that it brings out new physical laws.

Consider a common balance with a fixed counterpoise in the pan A. We place various bodies in the pan B. Some of these make B sink, others make it rise. Thus the balance with a standard counterpoise provides a way of classifying bodies. But there is something associated with the effect of the Earth on each body that is expressed by the quantity μg; μ being a property of the body. Since g is the same for all, the classification may be in terms of the values of μ.

We have a check on this. If we return to the problem of the Solar System and suppose that the distances between the bodies indicated by $\beta = 1, 2, \ldots, k$ are all small compared with their distances from the $(k+1)$th body, the acceleration of the $(k+1)$th body due to all together is nearly

$$(\mu_1 + \mu_2 + \ldots + \mu_k)/r^2$$

towards their centroid. Thus for particles close together the effects on a distant body are expressible by adding the corresponding values of μ. If we place several bodies in the pan together, the effect of the balance on all together is in opposition to the effects of the Earth on all together; and since we must measure the effects of all on the Earth by the sum of the values of μg, we naturally measure the effect of the Earth on all together by the same sum. If so, the mass measured by the balance, m, measured as an elementary magnitude, is proportional to the derived magnitude μ. Being an additive magnitude, mass has to be measured in terms of a unit; the value of μ per unit mass remains to be determined. We shall denote it by f, so that $\mu = fm$.

The important constant f has to be determined by direct observation of the attractive force between bodies of known mass at the Earth's surface. For weighable bodies this force is extremely small, but it can be measured. It is found that the couple needed to twist a fine fibre of vitreous quartz through any angle is proportional to that angle. A bar with two gold or lead spheres at its ends is suspended at its centre from such a fibre, and the period of the oscillation of the bar as it performs torsional oscillations is determined. The moment of inertia of the bar being known, this gives the couple exerted by the wire for any twist. Then the bar is allowed to take up its equilibrium position. Two large lead spheres are then arranged so that their attractions tend to twist the wire, and the head of the wire is then turned round until the torsion of the wire brings the bar back to its original position. The amount of turn needed determines the magnitude of the attraction and hence the constant f. The actual value is

$$f = (6 \cdot 670 \pm 0 \cdot 0064) \ 10^{-8} \ \text{cm.}^3 / \text{gm. sec.}^2, \ (3 \text{ degrees of freedom})$$

as calculated in my *Theory of Probability*, p. 281, from results of C. V. Boys and P. R. Heyl.

We can now extend the notion of mass to bodies too large or too small to be weighed on a balance, by saying that it is proportional to μ, which exists in general. Then our law 7·3 (14) relating to the effects of two bodies on each other takes the form, for any pair of masses m_1 and m_2,

$$m_1 \ddot{x}_{12} + m_2 \ddot{x}_{21} = 0.$$

In this form we call $m_1 \ddot{x}_{12}$ the force on m_1 due to m_2, and $m_2 \ddot{x}_{21}$ the force on m_2 due to m_1. Then we have Newton's third law in its usual form, that the forces on two bodies due to each other are equal and opposite.

We have also the law that the acceleration of a body is the sum of those due to other bodies in the world, obtained by adding the components in Cartesian co-ordinates. If we simply multiply this equation by the mass of the body, we have the rule for the composition of forces due to different bodies, that the force on a body is the resultant of the forces due to the other bodies, obtained by adding their components in Cartesian co-ordinates.

7·6. The reasons why force is interesting are then, first, that it has the symmetrical property that action and reaction are equal and opposite; secondly, that the forces on a body due to other bodies are additive; thirdly, that when the state of a system is known the forces are found to be determinate functions of the co-ordinates and possibly the velocities. Thus the equations of the form $m\ddot{x} = X$ are strictly differential equations for the co-ordinates.

Strictly these results have been established only for bodies moving freely under gravity. We proceed to apply them in general, even when the phenomena of impenetrability and friction arise. Their complete verification is then impossible, because motion in at least one co-ordinate is always prevented, and though there may be a force between the apparatus and the Earth we can never measure the corresponding acceleration of the Earth.

We can, however, notice that hanging a body on a spring produces an extension of the spring, approximately proportional to the mass measured on a balance. Though the spring comes to rest, its length is changed, and we have a verification that prevention of the acceleration of the body is associated with changes in other bodies; the notion of reaction preventing movement ceases to look so artificial.

Now consider a body with a plane face resting on an inclined plane at an angle α to the horizontal. It is subject to a vertical force mg, due to the Earth. This is equivalent to a force $mg\cos\alpha$ normal to the plane and another $mg\sin\alpha$ down the plane. If the body remains at rest, these must be balanced by reactions due to the plane, namely $N = mg\cos\alpha$ normal and upward, $F = mg\sin\alpha$ up the plane. We call these the normal and frictional reactions.

If the slope of the plane is gradually increased it is found that at a certain inclination λ the body can no longer remain at rest, but slides down the plane. At greater inclinations it has an acceleration approximately $g(\sin\alpha - \tan\lambda\cos\alpha)$. There is still no normal acceleration, so the normal reaction remains $mg\cos\alpha$. But the frictional reaction is $-mg\tan\lambda\cos\alpha$; thus we can express the data by saying that the ratio of frictional to normal reaction is never more than a constant $\tan \lambda$; if there is no slipping, the ratio takes up such a value, $\leqslant \tan \lambda$, as will suffice to prevent slipping.

But if the body is projected up the plane, or if it is projected down the plane when $\alpha < \lambda$, it is found that the portions $mg \sin \alpha$ and $-N \tan \lambda$ behave differently. The former always acts down the plane. The latter acts against the direction of motion, whatever that may be. This is the simplest example of the distinction between conservative and non-conservative forces. The work done by friction is $-N \tan \lambda$ times the length of the path, and is therefore greater in magnitude for a curved path than for a straight one, even with given termini. This work reappears as heat.

7·7. We can rewrite 7·3 (4) in the form

$$m_\alpha \ddot{x}_\alpha = \partial W / \partial x_\alpha, \text{ etc.,} \tag{1}$$

where

$$W = \sum_{\alpha \neq \beta} \frac{f m_\alpha m_\beta}{r_{\alpha\beta}} \tag{2}$$

It is convenient to make a change of notation at this point. We replace x, y, z by x_1, x_2, x_3. Thus x_α would be replaced by $x_{\alpha 1}$, and the set of equations of motion (1) will be rewritten

$$m_\alpha \ddot{x}_{\alpha i} = \partial W / \partial x_{\alpha i}, \tag{3}$$

where i takes the values 1, 2, 3 in turn. These lead as before to

$$T = \sum_i \sum_\alpha \tfrac{1}{2} m_\alpha \dot{x}_{\alpha i} \dot{x}_{\alpha i} = W + \text{constant}. \tag{4}$$

We call T the kinetic energy of the system, and W the work function. We shall also have occasion to put $V = -W$ and call V the potential energy.

If we write, more generally, $X_{\alpha i}$ for the force on particle α in the direction of x_i, we have

$$m_\alpha \ddot{x}_{\alpha i} = X_{\alpha i} \tag{5}$$

and

$$[T]_{t_0}^{t_1} = \int_{t=t_0}^{t_1} \Sigma \Sigma X_{\alpha i} dx_{\alpha i}. \tag{6}$$

The integral is called the work done on the system. In any case where $X_{\alpha i} = \partial W / \partial x_{\alpha i}$ and W is a function of the $x_{\alpha i}$ alone (i.e. not containing the velocities or the time explicitly) the right side is still $[W]_{t_0}^{t_1}$. Such systems are called *conservative*. They have the important property that the integral has the same value for given initial and final positions of the system, no matter how $x_{\alpha i}$ may vary on the way; they are not restricted to be solutions of (1). On the other hand (1) contain a great deal of information not included in (6).

An elastic string has a work function of different form. If the natural length is l_0 and the increase in length is x, the tension is $P = \lambda x$, where λ is a constant. In an increase of length dx, since the increase is opposed by the tension in the string, the work done is $-\lambda x dx$ and the work function (work done by the tension) is

$$W = -\tfrac{1}{2}\lambda x^2 = -\tfrac{1}{2}P^2/\lambda. \tag{7}$$

The equations of motion can be put into a form that depends as directly on fundamental notions as the conservation of energy. Suppose the co-ordinates of a particle of mass m moving according to the equations of motion to be x_i. The x_i, in any particular motion, are definite functions of the time. Take any three other functions of the time, δx_i, restricted only to being differentiable and small. Then the equations of motion imply

$$m\ddot{x}_i \delta x_i = X_i \delta x_i. \tag{8}$$

Add these equations for all particles of the system and integrate from $t = t_0$ to t_1. Then

$$\int_{t_0}^{t_1} \Sigma m\ddot{x}_i \delta x_i \, dt = \int_{t_0}^{t_1} X_i \delta x_i \, dt. \tag{9}$$

Now imagine the system to be moved from time t_0 to time t_1 in such a way that at time t the co-ordinates are $x_i + \delta x_i$. We may call $x_i + \delta x_i$ a varied co-ordinate and $\dot{x}_i + \delta \dot{x}_i$ a varied component of velocity. Then

$$\delta \dot{x}_i = (\dot{x}_i + \delta \dot{x}_i) - \dot{x}_i = \frac{d}{dt}(x_i + \delta x_i) - \frac{d}{dt}x_i = \frac{d}{dt}\delta x_i. \tag{10}$$

The left side of (9) is

$$[\Sigma m\dot{x}_i \delta x_i]_{t_0}^{t_1} - \int_{t_0}^{t_1} \Sigma m\dot{x}_i \frac{d}{dt}\delta x_i \, dt. \tag{11}$$

But $\qquad \dot{x}_i \dfrac{d}{dt}\delta x_i = \dot{x}_i \delta \dot{x}_i = \tfrac{1}{2}(\dot{x}_i + \delta \dot{x}_i)^2 - \dot{x}_i^2 - (\delta \dot{x}_i)^2;$ (12)

so that if $\qquad\qquad T = \Sigma \tfrac{1}{2}m\dot{x}_i^2,$ (13)

$$\Sigma m\dot{x}_i \frac{d}{dt}\delta x_i = \delta T - O(\delta \dot{x}_i)^2. \tag{14}$$

If δx_i vanish at times t_0 and t_1, so that the initial and final values of the co-ordinates are not varied, the first term in (11) is zero. Then

$$\int_{t_0}^{t_1} (\delta T + \Sigma X_i \delta x_i)\, dt = O(\delta \dot{x}_i)^2. \tag{15}$$

If the forces are conservative,

$$\Sigma X_i \delta x_i = \delta W + O(\delta x_i)^2, \tag{16}$$

and hence

$$\int_{t_0}^{t_1} \delta(T + W)\, dt = O(\delta x_i^2, \delta \dot{x}_i^2). \tag{17}$$

If we define a function

$$L = T + W, \tag{18}$$

and

$$S = \int_{t_0}^{t_1} L\, dt, \tag{19}$$

then, for small variations in the path, δS is of the second order in the variations. Further, if S is stationary for *all* variations δx_i that are differentiable and vanish at the termini, it can be proved that the equations of motion are satisfied. This is *Hamilton's principle.*

Notice that in general the co-ordinates $x_i + \delta x_i$ do not satisfy the equations of motion; the whole argument depends on considering motions that are imaginable, but physically impossible under the system of forces considered. Those who insist that all phenomena considered must be realizable in practice must reject Hamilton's principle as meaningless and deprive themselves of a valuable method.

Some of the forces may be internal reactions within a rigid body. A rigid body may be regarded as a limiting case of an arbitrarily stiff elastic one. A generalization of (7) will then show that the part of the work function that arises from these reactions is zero (corresponding to infinite λ). Hence for any varied motion that does not alter the distances between particles of a rigid body the reactions contribute nothing to the variation of the work function. The position of a rigid body is completely specified by 6 variables (three co-ordinates of a specified particle of the body and three rotations about it) and the differential equations for these variables contain no contribution from the internal reactions. This is d'Alembert's principle. It can be stated: the internal reactions in a rigid body form a system in equilibrium among themselves. It can be verified in the laboratory. The late Dr G. F. C. Searle had an experiment on a gyrostat which could be used to estimate gravity to about 1 part in 10^4. In astronomy the precession and nutation of the Earth's axis give a quantitative check.

The equality of action and reaction can be tested for elastic forces by experiments on collision. It is found that when two bodies collide the sums $m_1 \dot{x}_{1i} + m_2 \dot{x}_{2i}$ do not change. Since m_1 and m_2 are originally defined in terms of gravitational phenomena, this extends their scope to an entirely different type of force.

It is not absolutely obvious that the accelerations produced by a given gravitational field on bodies of different materials would be the same; though Mimas, a satellite of Saturn, has a density about a tenth of that of the Earth and shows no abnormality in its motion. However a terrestrial test can be made by means of the Eötvös balance. A horizontal bar is suspended by a fine wire and carries a vertical rod at one end. The instrument was designed originally to measure second derivatives of the gravitational potential. The masses to be compared in this experiment are placed at opposite ends of the bar. If they are of different materials the accelerations due to gravity might be different, but those due to the Earth's rotation are the same, and it is shown that the result would be a couple on the suspension wire. The most recent test in this way is by Roll, Krotkov and Dicke.† They compared Cu with $PbCl_2$, Au with Al. It might be expected that the effect, if any, would be greatest for greatest difference in atomic number. They find it to be at most of order 10^{-11} of the whole.

† P. G. Roll, R. Krotkov and R. H. Dicke, *Annals of Physics* **26**, 1964, 442–517.

CHAPTER VIII

LIGHT AND RELATIVITY

It did not last: the Devil, howling Ho!
Let Einstein be! restored the *status quo.*

J. C. SQUIRE

8·0. So far we have been concerned with the experimental basis of the two fundamental subjects, Euclidean mensuration and Newtonian mechanics. There is nothing controversial in the results; the object is to bring out essential features of the method of establishing them. In the twentieth century, however, both these subjects have been found to need slight alteration; in one respect, in fact, a more serious alteration than any modern one had been made much earlier, but was treated as of secondary importance.

We have seen that the truth of the equations of dynamical astronomy in the form

$$\ddot{x}_{\alpha i} = -\sum_{\beta} \mu_\alpha \frac{x_{\alpha i} - x_{\beta i}}{r_{\alpha\beta}^3}$$

requires that the axes shall have no rotation and that the origin shall have a uniform velocity with respect to the centroid of the Universe. It is not strictly necessary to introduce the centroid of the Universe, since we need consider only differences of the co-ordinates of the bodies. We have also supposed that at each instant of time each particle has definite co-ordinates with respect to these axes. There is, however, a serious difficulty about the meaning of the time in these equations. At first astronomers thought that it was the time of observation. But this was found to be incorrect by Römer. When the periods of revolution of Jupiter's satellites were found by observation of their eclipses when near opposition, the results were used to predict eclipses when the planet was situated otherwise; and errors ranging up to a quarter of an hour were found. Römer gave the correct explanation, namely that the time in the equations of dynamics is not the time of observation, but the time when the event under consideration actually happened, and that in a visual observation the time of observation is later because of the time taken by light to travel from the source to the observer. Jupiter in opposition is nearer to us than at other times, and therefore light takes a shorter time to travel to us. We always see the event later, but the delay is least in opposition. The variation corresponded to a time of travel across the Earth's orbit of 996 sec. A recent value is 998·022 ± 0·034 sec. When the times were corrected to allow for this the anomalies disappeared.

It will be seen, however, that Römer's explanation was really in two parts. First, it is assumed that the laws of motion for satellites near Jupiter are the same as for bodies near the Earth and for the motion of Jupiter about the Sun; secondly, that light has a finite velocity. The two are combined in the assumption that the time in the equations of motion is the time of observation reduced by the travel-time of light, and that it is when the time is modified in this way that the equations of motion are true. But this is not the end of the matter. The time of travel of light from Jupiter to the Earth would average about 40 min. If the time in the equations for Jupiter's satellites is not the same as the time of observation on the Earth, presumably that for the motion of Jupiter itself is different; and each satellite is perturbed by the others and a re-definition of t is needed for every satellite if we are to have the equations of motion in a form applicable to all.

The immediate consequence is that the laws of mensuration and gravity are approximations based on the assumption that the velocity of light is infinite, and we want to know how they are to be altered if this velocity is taken as finite. There is a complication because the distant bodies are all accelerated, whatever system we use, and we want to deal with complications one at a time if we can.

We therefore begin with a simplified model, in which all the bodies considered are mutually unaccelerated. This requires first, that parts of the same body are mutually unaccelerated, so that the body is treated as rigid; secondly, that the different bodies are sufficiently far apart for their mutual accelerations to be neglected. The first part keeps as much of mensuration as we can; the second is a restatement of Newton's first law.

In the first place we use the terrestrial frame, which satisfies the first condition reasonably well (we are not considering earthquakes or tidal deformations of the solid Earth).

Direct measurements of the velocity of light near the Earth's surface were carried out by Fizeau and Foucault. The methods depend on the principle of sending out an intermittent beam of light to a distant object, where it is reflected, and observing the time that elapses between the flashes going out and coming back. It was found, as expected, that the time was proportional to the distance travelled. With modern methods the velocity found is 299,792·5 ± 0·4 km./sec.† The result agrees with that found from Jupiter's satellites within the uncertainty of the latter. Also light *in vacuo*, or in a uniform medium, travels in straight lines; again we are not considering gravitational effects or diffraction. Our object at the present stage is simply to provide a consistent standard of comparison.

When the source of light has a velocity of its own, various possibilities arise with regard to the effect of this on the velocity of light. If light consisted of a stream of corpuscles, it might be expected that the velocity of the source would be added vectorially to the velocity of the emitted light. But if the velocity of light is a fundamental physical constant we might expect that when light gets away from the source it settles down to move with its standard velocity and forgets about its source.

There are, however, various ways of testing for effects that depend on the inverse square of the velocity of light. The most crucial is the Michelson and Morley experiment. Consider a moving system consisting of a source of light, with a mirror distant l away from it in the direction of the velocity. The source and the mirror are both moving with velocity v. If the velocity of light, c, is an absolute constant, light leaving the source has a velocity $c-v$ relative to the system and will overtake the mirror in time $l/(c-v)$. After reflexion it is moving with velocity c again, but toward the source, and the relative velocity is $c+v$. Hence it returns to the source in time $l/(c+v)$, and the total time taken is

$$\frac{l}{c-v}+\frac{l}{c+v}=\frac{2cl}{c^2-v^2}.$$

Now suppose that the direction of the mirror from the source is at right angles to the velocity of the system. Then if t is the total time of transit from the source to the mirror and back again, the source has travelled a distance vt, and if the light on return reaches the new

† L. Essen, *Endeavour*, **15**, 90, 1956.

position of the source it has a component of velocity v in the direction of motion of the latter. Hence its transverse velocity is $(c^2 - v^2)^{\frac{1}{2}}$. But the distance travelled transversely is $2l$. Hence the time taken is $2l/(c^2 - v^2)^{\frac{1}{2}}$. This is shorter than the time in the former case, in the ratio $(1 - v^2/c^2)^{\frac{1}{2}}$. If then we can arrange for the two specimens of light to leave the source at the same time and for the distances of the two mirrors to be equal, the light that has travelled transversely will arrive back first, and therefore in a different phase. If the difference of phase is great enough interference will take place.

The experiment was carried out by Michelson and Morley, and has since been repeated by several other investigators. The two specimens of light were produced by a mirror silvered to semi-transparency and inclined at $45°$ to the original beam from a lamp. Half the light went through to one mirror; the other half went transversely to the other mirror. On reflexion to the semi-transparent mirror, the latter transmitted half of one beam and reflected half the other, the resulting beams now travelling in the same direction. The distances were made nearly equal. The whole apparatus was then turned through a right angle. Thus the time of transit of one beam should be increased and that of the other diminished, and if interference did not occur in one case it should occur in the other.

It was actually found that the rotation of the apparatus through a right angle made no difference. If two waves took the same time to travel backwards and forwards with one setting of the apparatus, they did so again with any other setting. The velocity of the system in this experiment was the resultant of the velocity of the Earth in its orbit and that of the Sun relative, one supposes, to the centroid of the Universe. The latter may be supposed constant; the former is reversed every six months. It might perhaps happen that the two cancelled in one position of the Earth with respect to the Sun; but actually the result was the same at whatever time of the year the experiment was carried out.

The experiment is one of extreme delicacy, since the expected times of travel differ only by about 1 part in 10^8; special precautions are needed against elastic distortion of the apparatus when it is moved.

The null result of the experiment showed that there was something wrong with the premises—in fact had the experiment been done in the seventeenth century it would have been considered a convincing argument for the Ptolemaic system. The distances, for this purpose, are the distances between the points of reflexion of the light, as measured when there is no relative motion, and are therefore to be understood as in mensuration. Distance being taken in this sense, it appeared that the apparent velocity of light, measured as the distance travelled divided by the time of passage there and back, was the same in any direction. The expression of this result is that if (x, y, z) are the co-ordinates, with respect to an origin, of the place where a light-wave is at time t, then

$$\left(\frac{dx}{dt}\right)^2 + \left(\frac{dy}{dt}\right)^2 + \left(\frac{dz}{dt}\right)^2 = c^2, \tag{1}$$

irrespective of the motion of the origin and the directions of the axes. If we are to compare phenomena occurring near two widely separated objects we need a way of comparing standards of distance and time near them. There has been a great deal of confusion on this point. Identity of standards is generally assumed, but on inspection is very hard to interpret. However we can at least suppose that the laws of mensuration hold in both systems; that light in both has a unique velocity and travels in straight lines; that objects in uniform motion in one system appear to be in uniform motion in the other; and that an event occurring at one place at a specified place and time occurs at a specified place and time in the other. If then we take another origin and use analogous variables, indicated by accented letters, we shall also have

$$\left(\frac{dx'}{dt'}\right)^2 + \left(\frac{dy'}{dt'}\right)^2 + \left(\frac{dz'}{dt'}\right)^2 = c'^2. \tag{2}$$

If we write
$$\left. \begin{array}{l} ds^2 = c^2 dt^2 - dx^2 - dy^2 - dz^2, \\ ds'^2 = c'^2 dt'^2 - dx'^2 - dy'^2 - dz'^2, \end{array} \right\} \tag{3}$$

then if either of ds or ds' is zero, the other is also zero. It follows that for all positions and times

$$ds' = k\,ds, \tag{4}$$

where k may be a function of (x, y, z, t). When (x, y, z, t) are given, (x', y', z', t') are determinate, so that the variables referred to one system of reference are definite functions of those referred to the other. Thus k cannot involve the velocities.

Now suppose that the two systems of reference are such that a particle moving with uniform velocity in a straight line with respect to one does so with respect to the other. This implies that the two sets of equations

$$\frac{dx}{ds}, \frac{dy}{ds}, \frac{dz}{ds}, \frac{dt}{ds} = \text{constants}, \tag{5}$$

$$\frac{dx'}{ds'}, \frac{dy'}{ds'}, \frac{dz'}{ds'}, \frac{dt'}{ds'} = \text{constants}, \tag{6}$$

are equivalent. Now consider a particle to move from one given position at time t_0 to another at time t_1. Then if (x, y, z) are given as functions of t during the transit, $\int_{t_0}^{t_1} \frac{ds}{dt} dt$ has a definite value. If we choose a slightly different path, the change of this integral is $\delta \int ds$. We can show that the conditions (5) are just the conditions that $\delta \int ds$ shall be of the *second* order with respect to the changes of x, y, z. Thus the equations of motion of a particle under no forces are equivalent to the condition that $\delta \int ds$ is of the second order.

Similarly, from (6) they are equivalent to the statement that $\delta \int ds'$ is of the second order. Hence, if a path is such that $\delta \int ds$ is of the second order, so is $\delta \int k\,ds$.

It follows that k is constant. For if k depended on x, y, z, we could take a path near the original one, but always on the side of it where k is greater than on the actual path. Then $\delta \int ds$ would be of the second order, but $\delta \int k\,ds$ would be of the first order. If k depended on t, we could take the same path but alter the rate of travel so that t has slightly different values for the same x, y, z, and could arrange for

$$(dx/dt)^2 + (dy/dt)^2 + (dz/dt)^2$$

to be increased when k is large and decreased when k is small. Then we get a first order change in $\int k\,ds$. In either case the hypothesis that k is variable leads to a contradiction.

We actually take k unity. No actual value can be given for it without some means of comparing standards in the two systems. We are free, however, to choose the directions of the axes and the positions of the origins as convenient. We take the relative velocity of the origins to be along the axis of x, and then choose the origin of the second system to have co-ordinates (Vt, o, o) with respect to the first. The axis of x' is also along the direction of the relative velocity. We assume that lines perpendicular to this direction have the same length in both systems. An experimental method of test would be by means of an instrument called a cathetometer, which consists of a telescope mounted on a rod with its axis at right angles to the rod, and able to slide along the rod. The optical axis therefore always lies in a fixed direction, and differences in a co-ordinate parallel to the rod can be measured by finding the distance between the two positions of the telescope such that the objects are seen on the cross-wires. A rod placed parallel to the axis of y and moving with velocity V along the axis of x could be measured with this instrument. The accuracy attainable is insufficient to detect differences of order V^2/c^2, but the comparison is possible in principle. Then if we agree to take the same rod as a standard of length we can take the axes of y' and z' so that $y' = y$, $z' = z$. When comparison of lengths in directions perpendicular to the line of sight has been established, that for other directions follows. For both systems we can use rectangular Cartesian frames, and the standard of length in any direction in each is fixed by turning a scale about. Similarly a standard of time can be taken as that needed for light to travel along the standard scale and back. With these conventions c' becomes equal to c.

Our usual standard of time is specified by the rotation of the Earth. This obviously could not be adapted to give a standard on another body, for which the rate of rotation would be different. A pendulum clock is also unsuitable. In fact the only standards that could be suitable would be intrinsic vibrations such as those of atomic clocks, and our comparison essentially assumes that the ratio of the times of an atomic vibration and the passage of light along a standard scale fixed in the body is the same whichever body of the system it is attached to. This is not a convention but a natural hypothesis.

We can also write

$$\frac{dx'}{ds'} = \frac{1}{k}\frac{dx'}{ds} = \frac{1}{k}\left(\frac{dx}{ds}\frac{\partial x'}{\partial x} + \frac{dy}{ds}\frac{\partial x'}{\partial y} + \frac{dz}{ds}\frac{\partial x'}{\partial z} + \frac{dt}{ds}\frac{\partial x'}{\partial t}\right) \qquad (7)$$

with similar equations. Now for a particle in uniform motion dx/ds, dx'/ds' and similar expressions are constants, but these constants are different for different particles. This can be true in general only if $\partial x'/\partial x$, $\partial x'/\partial y$, etc. are also constants. Therefore x', y', z', t' are linear functions of x, y, z, t.

The relations between the co-ordinates must now be of the form (with additive constants)

$$\left.\begin{aligned} x' &= \beta(x - Vt), \\ y' &= y, \\ z' &= z, \\ t' &= \alpha x + \beta_1 y + \gamma z + \delta t. \end{aligned}\right\} \qquad (8)$$

Substituting in (4) we have the identity

$$c^2(\alpha\,dx + \beta_1\,dy + \gamma\,dz + \delta\,dt)^2 - \beta^2(dx - V\,dt)^2 - dy^2 - dz^2$$
$$= k^2(c^2\,dt^2 - dx^2 - dy^2 - dz^2). \qquad (9)$$

On equating coefficients of $dx\,dy$, $dx\,dz$, $dy\,dt$, $dz\,dt$ we have

$$\alpha\beta_1 = \alpha\gamma = \beta_1\delta = \gamma\delta = 0. \qquad (10)$$

If $\delta \neq 0$, $\beta_1 = \gamma = 0$. Then from the coefficients of dy^2 and dz^2,

$$k = \pm 1; \qquad (11)$$

and from those of dx^2, dt^2, $dx\,dt$

$$c^2\alpha^2 - \beta^2 = -1; \quad c^2\delta^2 - \beta^2 V^2 = c^2; \quad c^2\alpha\delta + \beta^2 V = 0. \qquad (12)$$

Eliminating β and V we have

$$(1 - \delta^2)(1 + \alpha^2 c^2) = -\alpha^2 c^2 \delta^2, \qquad (13)$$

whence $$\delta^2 = 1 + \alpha^2 c^2 = \beta^2 \qquad (14)$$

and then from the second of (12)

$$\beta^2 = (1 - V^2/c^2)^{-1}. \qquad (15)$$

If x and x' are to increase in the same direction we must take β positive, and can take also α and δ positive. Then

$$k = +1, \quad \beta = (1 - V^2/c^2)^{-\frac{1}{2}}, \tag{16}$$

$$x' = \beta(x - Vt), \quad y' = y, \quad z' = z, \quad t' = \beta(t - Vx/c^2). \tag{17}$$

If we solve these for x, y, z, t we get

$$x = \beta(x' + Vt'), \quad t = \beta(t' + Vx'/c^2). \tag{18}$$

Thus $-V$ is the velocity of the first origin with regard to the second.

The equations (17) were first given in full by H. A. Lorentz, who showed that a transformation of this type left the form of the equations of the electromagnetic field unaltered. In the hands of Einstein they became the basis of the special theory of relativity.

If in (10) we take $\delta = 0$ and either β_1 or $\gamma \neq 0$, we must have $\alpha = 0$. Then the coefficient of $dx\,dt$ gives $V = 0$, while those of dx^2 and dt^2 give

$$k^2 = \beta^2, \quad k^2c^2 = -\beta^2V^2.$$

These are inconsistent and this alternative needs no further consideration.

Most derivations of the transformation (17) from the equivalence of $ds = 0$ and $ds' = 0$ proceed at once to $ds' = ds$. This is unsatisfactory because it is not true even that the former equivalence implies that ds' is a constant multiple of ds. An example to the contrary is given by E. Cunningham.† An extra condition is needed to ensure that k is constant; the condition used here is that a uniform motion in one system is uniform in the other, which is natural if we are to apply the theory to dynamics. The further assumption that $k = 1$ requires that there should be some standard common to the two systems to establish a comparison of scales. This has been achieved here by supposing distances perpendicular to the relative motion of the origins to have equal measures in the two systems; this cannot be verified accurately, but could in principle. If $k \neq 1$ all intervals of distance and time in the second system would be multiplied by k. Angles and velocities would not be altered.

---

† *The Principle of Relativity* (1914), p. 89. The most general transformation that makes the equivalence hold is a combination of the above transformation with scale changes, reflexions and inversions in 4-dimensional space-time.

8·1. Now consider two events specified in the first system by (x_1, y_1, z_1, t_1), (x_2, y_2, z_2, t_2) and in the second by corresponding quantities with accents. Then

$$x_2' - x_1' = \beta\{x_2 - x_1 - V(t_2 - t_1)\}; \tag{1}$$

$$y_2' - y_1' = y_2 - y_1; \quad z_2' - z_1' = z_2 - z_1; \tag{2}$$

$$t_2' - t_1' = \beta\{t_2 - t_1 - V(x_2 - x_1)/c^2\}. \tag{3}$$

Suppose that the events are simultaneous in the second system, so that $t_1' = t_2'$. Then from (3)

$$t_2 - t_1 = V(x_2 - x_1)/c^2, \tag{4}$$

and on substitution in (1)

$$x_2' - x_1' = \beta(1 - V^2/c^2)(x_2 - x_1)$$
$$= (1 - V^2/c^2)^{\frac{1}{2}}(x_2 - x_1). \tag{5}$$

Hence $x_2' - x_1'$ is less than $x_2 - x_1$ in a definite ratio. If on the other hand $t_1 = t_2$, we shall find

$$x_2 - x_1 = (1 - V^2/c^2)^{\frac{1}{2}}(x_2' - x_1'). \tag{6}$$

Now when we measure a distance on a moving object we compare the positions of the ends of the object simultaneously with those of points with no motion relative to our axes. The equality of t_1' and t_2' is essential to the attribution of any meaning to $x_2' - x_1'$ in terms of distance when the co-ordinates themselves are varying with time. If we have an object whose length in the x direction in the first system is independent of the time in the first system, then its length in the second system is less than its length in the first in the ratio $(1 - V^2/c^2)^{\frac{1}{2}}$. This apparent contraction of a moving object in the direction of motion is known as the Fitzgerald contraction, and depends, as we see from (4), on the fact that two observers in relative motion differ in their ideas of what events are simultaneous on account of the finite velocity of light.

We can now identify the time of a distant event in terms of light signals. For if a mirror is at a fixed distance l, light takes a time l/c to reach it, and a further time l/c to return. Thus the time when reflexion occurs is the mean of those when the wave leaves the source and returns to it. This result is irrespective of the velocity of the source, and is not true on the older theory, where the times of the outgoing and returning waves were liable to differ, as we saw in discussing the Michelson-Morley experiment.

The chief difficulty that was felt in relation to the theory of relativity was precisely in connexion with the result that events that are simultaneous with respect to one observer are not simultaneous to another. But this difficulty really arises at a much earlier stage than the transformation 8·0(17). If the time means the time of observation, then the observations of Jupiter's satellites prove that the equations of dynamics are untrue, and conversely, if the equations of dynamics are true, then the time of an event is not the time of observation. Thus the whole conception of simultaneity required rediscussion from the start. In any case the times of different observers' observations of an event differ by quantities of the order of the differences of r/c, where r is the distance travelled by the light. In the new theory we have found a time of the event itself, which differs for different observers by quantities of the order of Vr/c^2. But V/c is in general small; thus the deviations between different observers from agreement about time-intervals are of the second order of small quantities instead of the first. The objection was in fact a straining at the gnat, while swallowing the camel without even noticing the existence of the larger animal.

8·2. Now consider a point moving with velocities (u, v, w) with respect to the system (x, y, z, t). Then its velocities with respect to the second system are

$$u' = \frac{dx'}{dt'} = \frac{\beta(dx - V\,dt)}{\beta(dt - V\,dx/c^2)} = \frac{u - V}{1 - uV/c^2}, \tag{1}$$

$$v' = \frac{dy'}{dt'} = \frac{dy}{\beta(dt - V\,dx/c^2)} = \frac{v}{\beta(1 - uV/c^2)}, \tag{2}$$

$$w' \qquad\qquad = \frac{w}{\beta(1 - uV/c^2)}. \tag{3}$$

We notice that if $(u, v, w) = (c, 0, 0)$, then $(u', v', w') = (c, 0, 0)$, whatever V may be. If $(u, v, w) = (0, c, 0)$, then $(u', v', w') = (-V, c/\beta, 0)$, and
$$u'^2 + v'^2 + w'^2 = V^2 + c^2(1 - V^2/c^2) = c^2.$$

These results are of course particular cases of our fundamental rule that the velocity of light is the same, however the observer is moving.

We notice a curious result if V should happen to be greater than c, the velocity of light. Imagine a particle moving with velocities (u, v, w), and consider its velocities with respect to an observer moving with velocity V. If $V > c$, β is imaginary. Hence v' and w' are imaginary, and the particle could have real co-ordinates at only one instant. For the rest of time y' and z' are imaginary. There seems to be no inherent contradiction in the idea of velocities greater than that of light; but if we consider as our universe all particles with velocities less than c with respect to ourselves, then any particle with velocity greater than c with respect to ourselves has a velocity greater than c with respect to any other particle of our universe. The world could then be classified into universes, such that no particle in any one universe could be perceptible for more than a fleeting instant from a different universe.

8·3. We now consider other observable consequences of the transformation. Consider a source of light sending out waves of period $2\pi/\gamma$ along the axis of x. Then the disturbance at any instant contains a factor such as

$$\phi = A \sin \gamma(t - x/c). \tag{1}$$

Consider an observer with a velocity V along the x axis. Using 8·1 (18) we have

$$\phi = A \sin \gamma\beta\left(t' + \frac{Vx'}{c^2} - \frac{x'}{c} - \frac{Vt'}{c}\right) = A \sin \gamma\beta\left(1 - \frac{V}{c}\right)\left(t' - \frac{x'}{c}\right)$$

$$= A \sin \gamma'(t' - x'/c), \tag{2}$$

where
$$\gamma' = \gamma\beta(1 - V/c), \tag{3}$$

so that the period of the disturbance reaching the observer is longer than that estimated by a stationary observer in the ratio

$$1/\{\beta(1 - V/c)\} = \{(c + V)/(c - V)\}^{\frac{1}{2}}. \tag{4}$$

In practice V/c is always small; but the result is an apparent increase of the wave-length of a given spectral line for a receding star, and a shortening of it for an approaching one. This is the Doppler effect, and is measurable. It leads to estimates of the radial velocities of stars, and in the case of double stars determines differences of radial velocity of the components, which are consistent with those inferred from the transverse movements when parallax determinations of distance are available.

8·31. Now suppose that an observer in the (x, y, z, t) system sees a star in the direction (l, m, n). Then the velocity components of the light from the star are

$$(u, v, w) = -(lc, mc, nc). \tag{1}$$

To an observer in the (x', y', z', t') system the apparent direction is (l', m', n'), and the velocity components are, by 8·2,

$$-l'c = u' = \frac{-lc - V}{1 + lV/c}; \quad -m'c = v' = -\frac{mc}{\beta(1 + lV/c)};$$

$$-n'c = w' = -\frac{nc}{\beta(1 + lV/c)}. \tag{2}$$

Thus $m'/n' = m/n$, and the directions (l, m, n), (l', m', n') lie in a plane including the axis of x. If

$$l = \cos\theta, \quad l' = \cos\theta', \tag{3}$$

$$l' - l = \frac{(1 - l^2)V/c}{1 + lV/c}, \tag{4}$$

or, if we neglect the square of V/c,

$$\theta' - \theta = -\frac{V}{c}\sin\theta. \tag{5}$$

Thus the apparent direction of the star is displaced away from the direction of the relative motion of the second observer by an amount given by (5). This is the phenomenon of *aberration*. On account of the Earth's orbital motion its velocity relative to the Sun varies in the course of a year, and therefore produces periodic variations in the apparent directions of the stars. These are the same for all stars in the same part of the sky, and are well known to astronomers. In fact, they give an estimate of the Earth's orbital velocity, and hence provide one of the data for the mean distance of the Sun.

It is easy to show, incidentally, that

$$l^2 + m^2 + n^2 = l'^2 + m'^2 + n'^2 = 1.$$

This is a check on consistency. But also, since the result (5) contains only ratios of displacements in different directions it is independent of the conventions $k = 1$ and $c = c'$.

8·32. Now consider light emitted from a source and entering water moving with velocity V in the direction of the beam. The velocity of light relative to the water is c/μ, where μ is the refractive index. The velocity of the source relative to the water is $-V$. Using 8·2 (1) to get the velocity of light relative to the source, we get

$$u' = \frac{c/\mu + V}{1 + (c/\mu)(V/c^2)} = \frac{c}{\mu} + V\left(1 - \frac{1}{\mu^2}\right) + O\left(\frac{V^2}{c^2}\right). \qquad (1)$$

This is tested in Fizeau's experiment. Water travels in a closed pipe so that when light is travelling outward to a distant pair of mirrors the water is moving with it, and the reflected beam travels with the return current, so that the effect of V is to increase u' on both journeys. Another beam is sent round the other way, so that its apparent velocity is reduced by the motion of the water. The two beams are recombined on return and the difference in the times of travel is measured by a method of interference. It was found in a repetition of the experiment by Michelson and Morley that the observed value of the coefficient of V was 0.44 ± 0.02.† The value calculated from the refractive index was 0.438; but this became 0.451 when a refinement allowing for dispersion was made. A further repetition by Zeeman gave still closer agreement.

The result of Fizeau's experiment is extremely important. If the velocity of light in a moving medium was the sum of the ordinary velocity and the velocity of the medium, the coefficient of V would have been 1. If the velocity was independent of that of the medium it would have been 0. The experiment excludes both these alternatives. It shows also that the actual coefficient agrees with that calculated from (1); and the term in $1/\mu^2$, if we trace it back, is found to come from the term in uV/c^2 in the denominator of 8·2 (1), which came in turn from the term in Vx/c^2 in the expression for t'. Thus it gives a direct check on this term, which is the very one in the fundamental transformation that was most subject to dispute. The

† A numerical correction due to Cunningham, *Relativity and the Electron Theory* (1920), has been used.

check is in fact more accurate than the corresponding one in the Michelson-Morley experiment.

The foregoing theory is usually known as Einstein's special theory of relativity, though the use of the word 'relativity', as if it expressed a novel feature, is really incorrect. The relativity of the equations of mensuration and dynamics, in the sense that they are true whatever unaccelerated non-rotating axes are used, was well known. The need for Einstein's theory arose from two facts about light: first, that it has a finite velocity; and second, that this velocity is independent of the motion of the observer. If light was propagated instantaneously we could identify the time of an event with that of observation, and there would have been no further trouble. The modern problem is not the discovery of relativity, but to retain relativity without introducing inconsistency with what we know about light.

A test for elasticity was carried out by A. B. Wood, G. A. Tomlinson and L. Essen.† They used two quartz rods with almost the same period of longitudinal vibration. Differences of period of a few parts in 10^{11} could be measured. As in the Michelson-Morley experiment the frame was rotated, so that an effect of the orientation with respect to the Earth's orbital motion could be sought. No change of period could be detected; if any, it was less than 1 per cent of the effect of the estimated contraction on the supposition that there was a change of length with no corresponding change of elastic moduli. Incidentally they mention (p. 610) five repetitions of the Michelson-Morley experiment, which all gave higher accuracy than the original one and still got a null result. It appears therefore that elastic vibrations could also be used to establish a comparison of standards of time.

When all electric charges and magnetization are fixed, potentials satisfying similar equations to those for gravity exist, and forces between them are derived analogously from the potentials. When they vary there is an interaction. An electric current produces a magnetic field, and a variable magnetic field produces an electric one. These interactions are utilized in dynamos and motors. They suggested to Maxwell and Hertz that the resulting equations should be applied to free space, and the result was that electromagnetic waves can exist. The predicted velocity of these waves was found

† *Proc. Roy. Soc.* A, **158**, 1937, 606–33.

to be that of light. (This is now verified directly to about 1 part in 10^5.) A charged particle in an electromagnetic field acquires an acceleration in the direction of the electric field and one at right angles to the magnetic field and to its velocity. This was verified for electrons.

It had been found by Larmor and Lorentz that Maxwell's equations have an invariance property; if two origins have a relative velocity and their (x, y, z, t) are related in the manner of special relativity, the equations remain true with a suitable relation also between the electric and magnetic fields. Einstein's contribution was to identify the fields so related with those actually observable by the two observers. The waves are those used in wireless telegraphy; visible light is a case of very short wave-lengths and X-rays of still shorter ones.

The wave theory of light was of course much older, but this work put it on a much firmer basis. Further, the rectilinear propagation of light was revealed to be an approximation for short wave-lengths; the edge of a shadow is not absolutely sharp. This is a case of *diffraction*. But the main thing is that the special theory of relativity connects many other phenomena than those considered in the elementary statement of it that we started with.

8·4. So far, however, we have treated only uniform motion. The equations of motion also need some modification. The special theory depends on three postulates: a particle moving with uniform velocity with respect to one inertial system of reference has uniform velocity with respect to any other inertial system; the velocity of light is the same in any such system; and for neighbouring events $ds = ds'$ for any two systems. The first two statements can be expressed by saying that the path of an unaccelerated particle is such that $\delta \int ds$ is of the second order for small variations of the path, the ends being kept fixed; and a light ray is a limit of such paths as $\int ds \to 0$ in every interval of the path. These have a very general form; and they must be a special case of motion in a gravitational field. Evidently ds is fundamental since all three conditions can be expressed in terms of it.

Now we saw that we could put the equations of dynamics in a very general form

$$\delta \int_{t_0}^{t_1} \{\Sigma \tfrac{1}{2}m(\dot{x}^2 + \dot{y}^2 + \dot{z}^2) + W\}\, dt = 0 \qquad (1)$$

to the first order. For particles under no forces $W = 0$. But

$$\int ds = \int \frac{ds}{dt}\, dt = c \int \{1 - \tfrac{1}{2}(\dot{x}^2 + \dot{y}^2 + \dot{z}^2)/c^2 + O(c^{-4})\}\, dt. \qquad (2)$$

If we do not vary the values of (x, y, z, t) at the ends of the path, the first term of (2) is just $c(t_1 - t_0)$ and its variation is zero. The second term, apart from a constant factor, leads to an equation of the same form as (1) takes for an unaccelerated particle. This strongly suggests that Hamilton's principle and the stationary property of $\int ds$ are expressions of the same rule. If in fact we consider

$$\int \Sigma m(c^2\, dt^2 - dx^2 - dy^2 - dz^2)^{\frac{1}{2}} - \frac{W}{c}\, dt, \qquad (3)$$

we have an integral that behaves in the proper way when W is zero, and yields Hamilton's principle as an approximation, with errors of order c^{-2}, when W is variable.

This holds for each body separately, and if we introduce U_α, the gravitational potential at the particle α due to the rest of the system, we can replace (3) by

$$\delta \int (c^2\, dt_\alpha^2 - dx_\alpha^2 - dy_\alpha^2 - dz_\alpha^2)^{\frac{1}{2}} - \frac{U_\alpha}{c}\, dt_\alpha = 0 \qquad (4)$$

for every α. But this integral is equal to

$$c \int \left\{1 - \frac{U_\alpha}{c^2} - \frac{\dot{x}_\alpha^2 + \dot{y}_\alpha^2 + \dot{z}_\alpha^2}{2c^2} + O(c^{-4})\right\} dt_\alpha,$$

so that if we redefine ds_α by the equation

$$ds_\alpha^2 = (c^2 - 2U_\alpha)\, dt_\alpha^2 - dx_\alpha^2 - dy_\alpha^2 - dz_\alpha^2 \qquad (5)$$

we can sum up our present knowledge in the form

$$\delta \int ds_\alpha = 0 \qquad (6)$$

for each particle, to order c^{-1}.

Clearly an infinite number of such hypotheses would satisfy our present data equally well; for all that is necessary is that the integrand should reduce, at a great distance from attracting bodies, to the ds of the special theory, and that near attracting bodies the second approximation should be of the form

$$c \int \Sigma m_\alpha \left\{ 1 - \frac{1}{2c^2} (\dot{x}_\alpha^2 + \dot{y}_\alpha^2 + \dot{z}_\alpha^2 + 2U) \right\} dt. \tag{7}$$

Ambiguity of t brings in terms of higher order in $1/c$.

Starting from our recognition of the fundamental importance of ds, we see from (5) and (6) that there is a possibility of retaining it in a gravitational field, provided that we alter the coefficients slightly. But if ds is to have such an importance it must be the same for all observers, and it is easy to see that this requires a corresponding modification of our whole system of Cartesian co-ordinates and time. Since $ds = 0$ for light rays, and gravity now appears in ds, light will be affected by gravity, light rays may be curved in a gravitational field, and our test of collinearity among distant objects is inaccurate; our laws relating distances between collinear objects also become approximations. We might try to save them by saying that in a gravitational field the ds suitable for light still has constant coefficients, so that light still travels in straight lines, but that the form suitable for material particles does involve the gravitational field. But at the present time this possibility is hardly worth discussing, because we do know that light rays are curved in a gravitational field.

What we are still entitled to say is that in any system of reference the position of a particle at any instant can be represented by three variables x_1, x_2, x_3, the instant being itself specified by a time-like variable x_0. Then we may say that an *event* is specified by the four variables x_0, x_1, x_2, x_3. If two events happen at neighbouring places at a short interval of time, we can say that ds^2 is a quadratic function of the four variables, the coefficients being functions of the variables. We write then

$$\begin{aligned} ds^2 &= g_{00} dx_0^2 + 2g_{01} dx_0 dx_1 + \ldots + g_{11} dx_1^2 + \ldots + g_{33} dx_3^2 \\ &= g_{ik} dx_i dx_k, \end{aligned} \tag{8}$$

where the g's are to be determined. In (8) we use the summation convention of tensor calculus, that where a suffix such as i or k is repeated it is to be given all possible values in turn and the results are added. By symmetry we can take

$$g_{ik} = g_{ki}. \tag{9}$$

In the absence of gravitation we can take x_0 to be ct and ix_1, ix_2, ix_3 to be the three Cartesian co-ordinates. Then

$$g_{00} = g_{11} = g_{22} = g_{33} = g_{44} = 1,$$
$$g_{01} = g_{12} = \ldots = 0, \tag{10}$$

or, briefly,
$$g_{ik} = \delta_{ik}, \tag{11}$$

where $\delta_{ik} = 1$ if the suffixes are equal and 0 if they are different.

In the presence of gravitation the g_{ik} will be modified. We have one consideration from Newtonian mechanics to guide us. The departure of the velocity of a particle from constancy depends on the first space derivatives of the gravitation potential U. Far away from matter, these are zero, and the particle moves with uniform velocity in a straight line. If they are constant, all particles have the same acceleration. Near other matter, these derivatives are not zero or constant; but a function formed from their variations with position, namely

$$\nabla^2 U = \frac{\partial^2 U}{\partial x^2} + \frac{\partial^2 U}{\partial y^2} + \frac{\partial^2 U}{\partial z^2}, \tag{12}$$

is still zero outside matter and proportional to the density inside it. Now it appears from (7) that the most important difference between g_{ik} and δ_{ik} is a multiple of U in g_{00}, and what we need is a suitable modification of (12) to give differential equations for the other g_{ik}, and possibly a more accurate one for g_{00}. It would be introducing complications prematurely to require these differential equations to be above the second order.

Einstein's procedure was to notice that the condition for absence of a field is equivalent to the condition that, by a suitable transformation of co-ordinates, ds^2 can be put into a form $g'_{jl} dx'_j dx'_l$ such that $g'_{jl} = \delta_{jl}$. With some forms of the original g_{ik} this is possible; with others it is not. When it is possible the system is called Galilean, and the special theory holds. The condition that it shall be possible is that a certain fourth order tensor B^l_{ijk}, depending on the g_{ik} and on their first and second derivatives with respect to the co-ordinates, shall be zero. This on the face of it has 256 components, but on

account of various symmetry relations only 20 are actually independent. The vanishing of all components of this tensor is the
condition for the absence of a gravitational field. In a Newtonian
system this would be $\partial^2 U/\partial x_i\,\partial x_k = 0$. Einstein then looked for a set
of equations formed from them corresponding to (12), and this was
naturally

$$R_{ik} = B_{ikm}^m = 0. \tag{13}$$

R_{ik} is a tensor of the second order, formed from the g_{ik} and containing only second and lower derivatives. Then (13) provides the
requisite number of differential equations for the g_{ik}. By the general
properties of tensors, if all components of R_{ik} vanish in one system
of reference they do so in all, so that we can write down these
equations in any co-ordinates we may choose. Inside matter R_{ik} is
not zero. When the field is not varying with the time, that is, if all
$\dfrac{\partial}{\partial x_0} g_{ik} = 0$, $R_{00} = 0$ is found to be equivalent to $\nabla^2 g_{00} = 0$, apart from
a term containing c^{-2}. Within matter, by analogy with Newton's
law, we may therefore say that R_{00}, like $\nabla^2 U$, is proportional to the
density.

The solution of the equations $R_{ik} = 0$ outside matter has actually
been carried out in only a few cases, and especially in that of
spherical symmetry. Eddington has remarked that the problem of
two bodies in general relativity is as difficult mathematically as that
of three bodies in ordinary dynamics. The spherical case is, however, the most important. We may imagine the time to be that of
an observer on the Sun, and the direction of a planet to be given by
the usual angular co-ordinates θ and ϕ. Another co-ordinate is
needed to specify the distance from the Sun. Now if we imagine
a short rod of length $d\sigma$ placed at right angles to the radius from the
Sun, it subtends a small angle $d\psi$ at the Sun. Then we take the distance r to mean $d\sigma/d\psi$. Then for such small displacements as make
dr and dt zero we have

$$ds^2 = -r^2(d\theta^2 + \sin^2\theta\,d\phi^2), \tag{14}$$

and in general

$$ds^2 = g_{00}(r)\,dt^2 + g_{11}(r)\,dr^2 - r^2(d\theta^2 + \sin^2\theta\,d\phi^2); \tag{15}$$

for by symmetry g_{00} and g_{11} must be functions of r only. Einstein
proceeded to obtain the R_{ik}, and found that they could vanish only if

$$g_{00} = c^2 g_4(r) = c^2(1 - 2fm/c^2 r); \quad g_{11} = -g_1(r) = -(1 - 2fm/c^2 r)^{-1}. \tag{16}$$

Here fm/c^2 is a length; it arises as an arbitrary constant in the solution of the differential equations, but comparison with Newton's theory identifies fm with the μ of dynamical astronomy.

The theory leads to three observable departures from Newtonian mechanics. The new law is found to imply that the path of a planet is not exactly an ellipse, but a slowly revolving ellipse, the direction of the major axis turning round at a constant rate. This change is inappreciable by observation except for Mercury, which was shown by Leverrier to have an outstanding departure from Newtonian theory of just this character; and the amount found from Einstein's theory agreed closely with that known to exist.†

Allowance for the effect of gravity on light showed that a ray would be curved near the Sun, so that stars would not be seen in quite their usual directions if the light from them passed near the Sun. The amount of the deflexion was calculated and was verified by two expeditions to observe the total eclipse of 1919.

The third point concerns the wave-length of a line in the spectrum of a body with a strong gravitational field. We have supposed that ds retains its value when an interval is transported, and this provides the only way in the theory of comparing distant standards. Now the vibration of an atomic source of light is such a standard, and we may expect that the increase of ds during a vibration is independent of position. Thus if T is the period, $g_{00} T^2$ should be a constant for a given atom when not in motion. But as the issuing light wave advances its period remains constant. Hence light from a body with a high gravitational potential should have a longer period than light from a similar atom on the Earth, and this would be shown by a displacement of spectral lines to the red. The effect was sought in the Sun and later in the companion of Sirius. For the Sun the results are still inconclusive because different lines show different displacements, as do those from different parts of the solar disk. These differences are of the same order of magnitude as the effect sought. The companion of Sirius is a very faint star but is shown by its spectrum to be hot at the surface. From the rate of total radiation at that temperature and the apparent luminosity the diameter was estimated as about that of Uranus; and the mass was known from the perturbations of Sirius to be about that of the Sun. The

† When the new calculations of the motion of Mars are completed a further check will be possible.

corresponding density is about 60,000 gm./cm.³. The result seemed incredible, but was tested by means of the Einstein spectral shift by W. S. Adams. This is about 30 times that in the Sun on account of the smaller radius. The test is difficult on account of the need to cut out the light of Sirius itself, but led to a good agreement. The explanation of the high density depends on properties of highly ionized matter and was given by R. H. Fowler on the basis of a theory of Fermi and Dirac.

The theory is therefore well supported by observation. It is complicated mathematically, though not so complicated as many other parts of mathematical physics. But it is a natural extension of more elementary ideas, and it is certain that any other theory that succeeded in combining gravitational theory and the properties of light into one system would be far more complicated. However, there are still some difficulties. The excess motion of the perihelion of Mercury, as a result of a recent rediscussion by G. M. Clemence,† is $42.66'' \pm 0.43''$ per century; the theoretical value is $43.03'' \pm 0.03''$, and nothing could be more satisfactory. For the displacement of star images, the theoretical value when the ray grazes the Sun's limb is $1.75''$. The first two expeditions‡ gave $1.98'' \pm 0.18''$ and $1.61'' \pm 0.45''$; but later results have been systematically larger, and there is a suggestion that the correct value is about $2.2''$. The spectral shift in the companion of Sirius, given by what seemed to be the most suitable set of spectral lines, and corrected for Doppler effect,§ is 22 ± 2 km./sec., as against the theoretical value of 20 km./sec. But other lines give much worse agreement, and as for the Sun there seem to be further complications.‖ It is noteworthy that these two phenomena are more concerned with interaction between gravitation and light than the perihelion shift, and it is perhaps not altogether out of the question that there may be some unfore-

† *Astr. J.* **19**, 1947, 361–64; *Proc. Amer. phil. Soc.* **93**, 1949, 532–34. I have revised his calculation slightly to take account of Rabe's redetermination of some of the planetary masses.

‡ F. J. Dyson, A. S. Eddington and C. R. Davidson, *Phil. Trans. A*, **220**, 1920, 291–333. Probable errors converted into standard errors.

§ Recalculated from Eddington, *The Internal Constitution of the Stars* (1926), p. 173.

‖ M. G. Adam, *M.N.R.A.S.* **108**, 1948, 446–64. E. Finlay-Freundlich, *Phil. Mag.* (7), **45**, 1954, 303–19; he remarks that *B*-type stars show 18 times the expected red shift, and that the theoretical value for the companion of Sirius is about 80 km./sec.

seen complication. The proper form of the equations of electromagnetic waves in a gravitational field is still uncertain, and the spectral shift also involves quantum theory.

The spectral shift has been tested more accurately by what is called the Mossbauer effect. An astonishingly sharp radiation is obtained such that it is possible to detect the shift between two floors of the same building. Determinations have been made by R. V. Pound and G. A. Rebka[†] and by Pound and J. L. Snyder.[‡] The latter used a vertical path of 75 feet and the effect was 0.9990 ± 0.0076 of the predicted one. The radiation consists of γ rays from atomic nuclei in solids, which give no recoil and no broadening from thermal agitation.

G. M. Clemence and V. Szebehely[§] show that the differences arising from the eccentricity of the Earth's orbit could produce annual differences in time keeping of an atomic clock of ± 0.0017 sec.

L. Silberstein mentioned a long time ago, when the effect was doubtful, that, since this effect arises from the dt^2 term in ds^2, if it was not present not even Newtonian gravity would exist. In view of the test by the Mossbauer effect, however, this does not call for a drastic change of the theory; we are now entitled to say that planetary motions establish the value of the term still more accurately than the latest experiments do.

Before the publication of Einstein's theory various attempts were made to account for the motion of the perihelion of Mercury on Newtonian lines. An interior planet was suggested, but it was never found. But there is matter within the orbit of Mercury, some forming the solar corona, and some reflecting the zodiacal light. Theories were in existence that appeared to account for the anomaly of the movement of Mercury by the attraction of this matter, and also for an anomaly that exists in the motion of the plane of the orbit of Venus.[||] The latter is not touched by Einstein's theory, but is small and might be due to observational error or to incompleteness of the calculation. Such matter might also account for the eclipse displacement by its refraction. If such matter existed in such quantity as to account for any important fraction of the motion of the perihelion

[†] *Phys. Rev. Letters* **4**, 1960, 337.   [‡] *Phys. Rev.* **140**, 1965, B788.
[§] *Astr. J.* **72**, 1967, 1324–6.

[||] Jeffreys, *M.N.R.A.S.* **77**, 1916, 112–18, discussing hypotheses of Seeliger and de Sitter.

of Mercury or the displacement of star images, the remainder would not be in accordance with Einstein's theory, which would therefore be false. But its amount can be estimated from the amount of light that it reflects, and appears to be much too small to account for any appreciable fraction of the observed effects.[†] These must therefore be due to a departure of the law of gravitation from Newton's form.

Newcomb's discussion of the secular perturbations of the inner planets made the discrepancy in the motion of the node of Venus about four times the standard error. However a discussion by R. L. Duncombe,[‡] after revision of some of the adopted constants, reduces this to $0.04'' \pm 0.13''$ (p.e) in a century, so the discrepancy has disappeared.

8·41. *The clock paradox.* There has been much discussion of this in the special theory. It arose originally in a popular work of Einstein. It is as follows. A body A ejects a much smaller body B with velocity V; this after travelling a certain distance rebounds from another body C and returns to A. A and B both carry clocks, which are synchronized before the ejection. Then according to A's system B's clock rate is $\sqrt{(1 - V^2/c^2)}$ times A's, and when B returns his clock will appear to have lost time. According to B's, A's will have lost time, and there is a contradiction. However, the argument supposes that there are no observers but A and B. Any other observer could detect the impulsive changes in B's motion, and consequently B is not in the class of bodies that the special theory could be expected to apply to. *The special theory, properly understood, refers only to unaccelerated observers.*

Since B's clock is compared with A's before departure, k is then 1. But the only obvious method of comparing standards after departure is by measures of transverse displacements. Suppose that the linear dimensions of B are of order a and its density ρ. If it acquires a velocity V in travelling a distance a, momentum

† Jeffreys, *M.N.R.A.S.* **80**, 1919, 138–54. It should be remarked that the theory of scattering used was very elementary and needs revision. However, in spite of the importance of the point, it has received no further attention, nor is my result mentioned in any standard book on relativity that I have seen. It is perhaps not out of the question that refraction might explain the slight excess of the eclipse displacement above the theoretical value.

‡ *Astron. J.* **61**, 1956, 266–8.

$\rho a V$ per unit area is acquired in time a/V, and the stress needed is of order ρV^2. Before expulsion this is balanced by reactions, during expulsion it is not. Then the stress implies changes of strain of order $\rho V^2/\mu$, where μ is an elastic constant of the material. If α is a velocity of elastic waves, α^2 is of order μ/ρ; hence the strains are of order V^2/α^2. But α for actual materials is much less than c. Hence the expulsion gives a ratio of dimensions differing from 1 by far more than $\sqrt{(1 - V^2/c^2)}$ does.

The clock paradox in fact arises entirely from inadequate attention to methods of comparing standards, and in particular to the fact that A and B may not be the only observers.† A's time is right (apart from neglect of the recoil, which would be very small if B is much less massive than A); B's is almost certainly wrong by more than the special theory says.

A full treatment would require the equations of the general theory, account being taken of elasticity. The full equations for elasticity on the general theory have not yet been given, but one important check has been given by A. R. Curtis.‡ In ordinary elasticity the velocity of sound in an incompressible material would be infinite; this contradicts the principle that no velocity can exceed that of light. Curtis pays special attention to the meaning of incompressibility under general relativity, and finds that the velocity of sound even in this extreme case cannot exceed that of light.

A case where comparison is possible, given general relativity, is studied by F. I. Mikhail.§ He considered two particles in gravitational orbits about a central mass, one in a circular orbit and the other projected radially and returning. With suitable initial conditions they can meet twice. It is found that there is a difference, approximately 3 times what is given by a naïve application of special relativity.

8·5. It is worth while to ask whether, given the excess motion of the perihelion of Mercury and the displacement of star images, any other law than Einstein's would account for them. A partial answer

† See G. Builder, *Austr. J. Phys.* **10**, 1957, 246–62; H. Dingle, *Austr. J. Phys.* **10**, 1957, 418–23; H. Jeffreys, *Austr. J. Phys.* **11**, 1958, 583–6.
  ‡ *Proc. Roy. Soc.* A, **200**, 1950, 248–61.
  § *Proc. Camb. Phil. Soc.* **48**, 1952, 608–15.

to this can also be given. If we return to 8·4 (15) and assume $g_4(r)$ and $g_1(r)$ expanded in powers of $1/r$,

$$g_4(r) = 1 + A_0/r + B_0/r^2 + \dots, \tag{1}$$

$$g_1(r) = 1 + A_1/r + B_1/r^2 + \dots, \tag{2}$$

then the equation $\delta \int ds = 0$ is equivalent to

$$\delta \int \frac{ds}{dt} dt = \delta \int L dt = 0, \tag{3}$$

where $\qquad L^2 = c^2 g_0(r) - g_1(r) \dot{r}^2 - r^2 (\dot{\theta}^2 + \sin^2 \theta \dot{\phi}^2). \tag{4}$

This leads by the methods of the calculus of variations to equations of the form of Lagrange's. The result is as follows.†

From the mean motions of the planets

$$A_0 = -2fm/c^2. \tag{5}$$

From the perihelion of Mercury,

$$A_1 - 2B_0/A_0 = 2fm/c^2. \tag{6}$$

From the eclipse displacement

$$A_1 - A_0 = 4fm/c^2. \tag{7}$$

This leads to $\qquad A_1 = 2fm/c^2, \quad B_0 = 0, \tag{8}$

and therefore

$$ds^2 = c^2 \left\{ 1 - \frac{2fm}{cr} + O\left(\frac{2fm}{cr}\right)^3 \right\} dt^2 - \left\{ 1 + \frac{2fm}{c^2 r} + O\left(\frac{2fm}{c^2 r}\right)^2 \right\} dr^2$$

$$- r^2(d\theta^2 + \sin^2 \theta \, d\phi^2). \tag{9}$$

Thus all the terms in Einstein's solution capable of producing observable effects are directly demonstrated by the observational data. The argument would, however, need correction if, for instance, the 4 in (7) should have to be replaced by 5. Since $2fm/c^2 r$ is of the order of 10^{-7} even for Mercury, any effect of higher terms must be utterly negligible unless they have remarkably large numerical coefficients.

One attempt to account for the excess motion of the perihelion of Mercury had been to make a small change in the index in the law

† This solution was originally given in *M.N.R.A.S.* **80**, 1919, 138–54, and repeated in the first edition of this book. I have not thought it worth while to rewrite it in full.

of gravitation. It would, however, have been more natural to include an extra term in the potential. It would then have been found that to account for the perihelion of Mercury the new term in $1/r^2$ would have to be of the order of v^2/c^2 times the first, v being the orbital velocity of Mercury. The possibility of a gravitational effect on light had been suggested by Newton and later by Laplace. The suggestion had been forgotten; but a discovery of this sort would certainly have revived interest and led to an experimental test. It would then have been found that the actual deflexion of light in an eclipse is twice what Newton predicted, and it would have been seen that a more drastic revision of Newton's law was needed than the mere addition of a cube term to the gravitational acceleration.

8·6. Einstein also attempted to apply the general theory to the universe as a whole. The equations of motion require some hypothesis on the relation of stress to velocity, and this is the most doubtful part of the theory. He introduced an extra term into the equations outside matter, giving

$$R_{ik} = \lambda g_{ik}.$$

λ is a sort of world curvature. The treatment depends on the boundary condition at large distances. It would for instance be possible to assume that at a suitably large distance the mass outside that distance tends to zero and so does the outward flux of radiation (that is, all radiation is attracted back). In de Sitter's version the light from distant objects should show a spectral shift to the red; this is absent from Einstein's. Einstein originally thought that there was a reason why $\lambda \neq 0$, but in later papers (one with de Sitter) he declared himself no longer convinced of the necessity; the gravitation of actual matter might be enough by itself. Eddington continued to think that there was such a reason, on the epistemological ground that atomic processes provide one scale of length, and some other standard is needed so that one can be measured against the other. I see no reason whatever for this. (See under Mach's principle, p. 182.) At the time there was some sign of a systematic shift, but some objects showed the opposite and no general conclusion could be drawn. Since then, however, much material on the spectra of distant nebulae has been gathered, and there is now no doubt about the existence of the effect. If it was interpreted as a Doppler shift it would imply that the velocities of the distant nebulae are nearly proportional to their distances, as if they had all been close together about 3000 million years ago. The effect is very large; the change of wave-length reaches about 30 per cent.

Einstein's solution referred to a universe in a steady state. In a sense it is of finite size, but this is a matter depending on the definition of distance that is used. It is certainly of finite mass. On this point we may refer to an old argument of Olbers,[†] recently re-discovered by H. Bondi and T. Gold.[‡] Suppose that the Universe had Newtonian properties, was of infinite extent, and similar in all regions. Then the mass within distance R is proportional to R^3 and that in a shell of thickness dR to $R^2 dR$. As seen from the Solar System, since the intensity of light falls off as the inverse square, the light from such a shell would be of intensity proportional to dR. Thus in an infinite universe the total intensity of light would be infinite. Actually in such conditions the bodies would partially hide one another, but the effect would be that the whole sky would glow with the intensity of the Sun's disk. We know now that there is a great deal of dark matter between the stars, and it may be suggested that this absorbs most of the radiation. But this is no answer, because absorption of radiation would raise the whole of the dark matter to stellar temperatures. The Olbers paradox could be re-solved if either (1) the reddening of distant stars becomes so strong as to bring their radiation into the infra-red or (2) if no R can be greater than cT.

The idea of a Universe in a permanent steady state might have appeared fairly satisfactory, but Lemaître showed that the Einstein universe is actually unstable and must proceed either to collapse or to expand. Thus at least part of the apparent expansion of the Universe is genuine, and we have to face the question: did the Universe come into existence about 3000 million years ago? From evidence from radioactivity we are led to a similar estimate of the ages of the Earth's crust and of meteorites. On the other hand recent papers by Bondi and Gold and by Hoyle have shown that the data would be fitted equally well if new matter is continually

† *Bode's Jahrbuch*, 1826, p. 10.   ‡ *M.N.R.A.S.* **108**, 1948, 258.

being created. In these theories, the expansion keeps pace with the creation in such a way that the local density remains always the same.

This seems to be a case where scientific method is stopped for lack of data. More and more work is being done on these problems, but there is still no definite evidence of any effect of the scale of the universe except for radial velocities proportional to the distance.

Several hypotheses fit the same data, as far as we have any, and any judgment between them seems to be a matter of personal preference. Some people are horrified by the idea of instantaneous creation, others by that of continuous creation, and, so far as I can see, there is no more to be said. There is at present no explanation of either.

What happens if there are gravitational and electromagnetic forces simultaneously? The special theory already deals with the latter. But they interact in the theories of the eclipse displacement and the spectral shift. There have been at least three attempts to state equations that combine them, by Weyl, Eddington, and Einstein. So far no definite decision has been made between them or possibly some other. In cases where either effect is appreciable the other is small, and the effect of their product is far below the possibility of observation.

A 'ballistic' theory of light emission was proposed by W. Ritz (probably not for the first time) according to which light travels with its standard velocity combined with the velocity of the source. I pointed out in the first edition of this book that in the case of an eclipsing binary star such as Algol this would make the difference in transit times, according as the bright component is approaching or receding, exceed the orbital half-period, and the whole light curve of the star would be upset. I deleted this from the second edition because the velocities themselves were determined from the Doppler effect, and I could not see how to correct for this. However, later work has compared the aberration deduced from near stars with that from distant nebulae. On the theory in question the major axis of the aberration ellipse would differ from that for a stationary object by a factor $(c/c') - 1$, where c is the ordinary velocity of light and c' is that from the moving object. This has been tested by R. J. Dickens and S. R. C. Martin.[†] Distant

† *Observatory* **85**, 1965, 260–2.

nebulae with apparent radial velocities of 20,000 to 110,000 km./sec. were compared with galactic stars. The difference was not significant and was at the most a few per cent of that predicted. Other comments are made by R. Griffin. †

However, Fizeau's experiment already shows that motion of the whole medium is not simply added to the velocity of light. It would have been most surprising if motion of the source alone was.

8·7. Some 'philosophical' points, however, do seem to need comment. In Eddington's accounts gravitation was spoken of as a property of 'space'—or rather of the four-dimensional space-time. Now space had three meanings. To the older physicists space was just nothingness. To those of Larmor's generation it was filled with a mysterious 'aether', invented so that its vibrations would satisfy the equations of electromagnetic waves. In pure mathematics a space is simply the set of values that a variable or set of variables can take. Space in the first sense could not affect the motion of matter. Many physicists pointed out that what matters is the electromagnetic equations and the consequences that can be drawn from them; the 'aether' was a hypothetical entity with special properties chosen only to provide an explanation of these equations, and could perfectly well be done without. In the mathematical sense, however, properties of a space of measures are simply relations between measures. But this amounts only to saying that there is such a subject as dynamics. Eddington insisted that the geometrical interpretation was something different from the dynamical one, but apparently this only means that ds, which is analogous to an element of distance in four-dimensional geometry, plays a dominant part. But even for the same metric, ds^2 could be defined in an infinite number of different ways, so far as mathematics is concerned. The essential point physically is that ds plays three different parts in the theory, and further that the g_{ik} in it satisfy a particular set of differential equations; we could presumably always define a ds to satisfy one of the conditions, but, having defined it, we cannot be sure without physical test that it will satisfy any of the others. At any point the experimental results might have been different, and if so one of the

† *Observatory* **85**, 1965, 264–5.

hypotheses would have had to be modified. We can admire the ingenuity that led to the theory, but there was no inherent necessity that it must be true; its acceptance depends on observation.

Milne's kinematic relativity starts with a set of observers in relative motion, who can exchange information by means of light signals. He shows that in certain conditions they can graduate their clocks so that each develops a set of co-ordinates and time in which every other has a place; but the transformation reached connecting the graduations is that of the special theory of relativity. Like Eddington, he claims that this arises only from the principle that measurement is possible; but since he provides no place for the form of the ds^2 of general relativity, with variable g_{ik}, he has evidently introduced some additional restriction. In fact his form implies no effect of gravitation on light, and is in direct contradiction to the eclipse displacement. He further assumes that the universe looks alike to all observers wherever situated; this leads to a finite universe expanding with the velocity of light, and any observer would estimate the density as tending to infinity at the boundary. Now this is an entirely new principle with no observational support whatever. The first point of physics is that observations vary, both with regard to time and place. But what the new principle does is to say, not only that there are general laws of motion, but that the initial conditions also are subject to severe restriction. With much effort the laws of motion, which are differential equations, seem to have been found; and they have the property of being capable of being stated in a form that is the same for all observers. No such property can be inferred for initial conditions, since the laws are applicable to any initial conditions.

Milne gets an approximation to the inverse square law of gravitation from his theory, but he has not explained the motion of the perihelion of Mercury, which must be a primary requirement for any satisfactory theory.

The same 'cosmological principle' is used by Bondi and Gold in their theory, which does take account of gravitation and leads to the conclusion that to maintain the cosmological principle it is necessary to have continuous creation of matter. As indicated earlier, I think their answer may be right; but the principle itself is baseless.

One good feature of Milne's work and that of Bondi and Gold is the insistence on precise statement of the methods proposed for actual measurement of the space-time co-ordinates. The equations of general relativity being meant to hold in all co-ordinate systems, there is no need for such specification in developing the theory itself. But it is needed for actual application, and it must be said that the specifications are left very vague in most statements of general relativity. In the treatment of the field of the Sun in 8·4 I have indicated how r is to be interpreted if ds^2 takes the form (15). In the problem of the expanding universe, however, there is no question of measuring the distance of a distant nebula by comparing its angular diameters at different times. What is actually done is to assume that the intrinsic luminosities of nebulae are distributed about a mean according to the same law whatever their distance may be; then the apparent brightness of a nebula can be used to estimate its distance by comparison with others that are near enough to have their distances estimated by other methods. It is not obvious, however, that this estimate will be the same as would be found by the angular diameter method if that were applicable. (The remote nebulae in a photograph look like small stars and there does not seem to be any hope of measuring their angular diameters and applying the statistical method to them.)

8·8. It is desirable to close this chapter with some reference to various dilemmas that are held to have some relevance to the theory of relativity. One that is not often mentioned but perhaps brings out the point most clearly is a comparison between Alice and Gulliver. Alice found herself growing larger or smaller, and assumed no change of size among her companions. Gulliver found himself among people far larger or far smaller than himself, and assumed no change of his own size. Their data were essentially similar; what they observed was a change of the ratios of their own sizes to other people's. Alice took the other people's sizes as a standard; Gulliver took his own.

Poincaré considered a spherical world where all objects contracted radially as they approached the boundary, in such a way that a given scale would have to be transported end to end an infinite number of times to reach the boundary. He argued that in such a world our procedure of measurement would lead to exactly the same rules as we get; thus there is a fundamental indeterminacy in our knowledge of the world, which might be really very different from what we think it to be.

Clifford† considered a worm in a closed tube, able to travel in only one direction around the tube. In a circular tube its experience could never change and it could regard the universe as one-dimensional and infinite. If the tube was not of uniform curvature, the worm's sensations would vary, and it could attribute this either to a change in the character of space or to a change in its own physical condition.

Such arguments had a great deal to do with the formulation of the principle of general relativity, which insists that laws must be stated in a form that is valid for all systems of reference and denies that the problems set by Clifford and Poincaré had any meaning. The odd thing is that anybody should have thought that verifiable results could follow from such a principle. All that is being said is that the laws can be expressed in terms of different systems of reference, and if relations exist for expressing one set of co-ordinates in terms of another, any equation that holds in one system can be transformed into an equivalent equation in the other. It is maintained that no system should be privileged, which means, if it means anything, that the laws must not take a specially simple form in any particular system. But at the outset this is false; the laws of mensuration and dynamics do take specially simple forms in terms of rectangular co-ordinates and time. If there could be a system with these properties near every event, we should naturally expect it to be privileged in this sense. There was never any *logical* reason for abandoning this hope.

Further, actual procedure still creates a privileged system. The excess motion of the perihelion of Mercury depends on a reference to non-rotating axes. As the stars have different proper motions the axes were originally taken so that faint stars had no systematic motion with respect to them. Most observed stars, however, belong to our own galaxy. Oort compared the proper motions of galactic and extra-galactic stars and found that the galaxy itself has a general rotation with respect to extra-galactic bodies. This rotation can be estimated in another way, on the hypothesis that the rotation of the galaxy is in its own plane. On Newtonian dynamics, within

† *Common Sense of the Exact Sciences* (1885), 214–26.

the domain where it agrees with Einstein's, the plane of the resultant angular momentum of the Solar System is fixed. The mean plane of the galaxy is strongly inclined to it. Thus the apparent motion of the former plane with respect to galactic stars gives another estimate of the galactic rotation, and the two estimates have been shown by Clemence to be consistent. Hence extra-galactic stars can be held to specify a non-rotating frame of reference. We could if we liked transform the equations so as to refer to a frame with an arbitrary rotation with respect to extra-galactic stars, but that does not alter the fact that the equations are more manageable if we do not.

The constructive achievement of Einstein was to combine what was known of gravitation and light into a single scheme and to find the appropriate generalization of Laplace's equation when the finite velocity of light is taken into account. This was a great feat of scientific imagination, and I think that it would be more appreciated if the bad philosophy associated with the theory was cleared away so that the actual achievement can be seen. The 'general covariance' of the laws is a platitude.

The restriction to unaccelerated observers leaves it possible for an unaccelerated observer to observe an accelerated body. A possible application is to a hydrogen atom. If we transform the masses according to the rule

$$m_1 + m_2 = \mu_0, \quad \frac{m_1 m_2}{m_1 + m_2} = \mu_1$$

and the co-ordinates by

$$m_1 x_{i1} + m_2 x_{i2} = \mu_0 \xi_{i0}, \quad x_{i2} - x_{i1} = \xi_{i1}$$

the kinetic energy takes the form

$$\tfrac{1}{2}(\mu_0 \dot{\xi}_{i0}^2 + \mu_1 \dot{\xi}_{i1}^2).$$

If there is no external force $\dot{\xi}_{i0}$ will be constant, and we can imagine an unaccelerated frame based on a particle at ξ_{i0}, what Eddington calls the extracule. The distance between the particles, r, would be measured in this frame, and, as Eddington appears to say, is not subject to relativity transformations. A related point in the general theory would be that in the motion of the Sun and a planet, r can still be defined as the distance between them (measured in a suitable way); but the full treatment of the problem of two gravitating bodies on the general theory has not yet been worked out.

8·9. *Mach's principle.* The equations of dynamics have been stated above as for non-rotating axes, which we call inertial axes. If instead we use rotating axes, they are altered by the introduction of terms in the products of the rate of rotation and the rates of change of the co-ordinates. But the stars as a whole have little or no systematic variation of direction with respect to the inertial axes. Is there any reason why this should be so? Mach suggested that inertia itself is a consequence of the totality of distant matter. This, on the face of it, is an example of the principle of causality, which Mach himself had taken a leading part in undermining, by insisting on science as the description of facts. In this aspect there is no more to do than to note the fact. On the other hand a mystery is made of the equivalence of gravitational and inertial mass. Now in the treatment of dynamics in Chapter 7 we saw that one quantity, μ, characterizes the dynamical properties of the body, and could be regarded as a derived magnitude. In fact T and W in 7·7, if multiplied by f, become equations containing μ; reference to m disappears. The Eötvös experiment comparing different materials is a check—it shows that we do not need this complication.

A discussion of various aspects of Mach's principle is by Hoyle and Narlikar,† which gives references.

† F. Hoyle and J. V. Narlikar, *Proc. Roy. Soc.* A, **273**, 1963, 1–11.

CHAPTER IX

MISCELLANEOUS QUESTIONS

'How is bread made?'
'I know that!' Alice answered eagerly. 'You take some flour—'
'Where do you pick the flower?' the White Queen asked. 'In a garden, or in the hedges?'
'Well, it isn't picked at all', Alice explained. 'It's *ground*—'
'How many acres of ground?' said the White Queen. 'You mustn't leave out so many things.'

LEWIS CARROLL, *Through the Looking Glass*

9·0. Three beliefs pervade most modern accounts of scientific principles and are the principal source of confusion. The first is that in some sense scientific laws are statements made with certainty. The second is that physical measures can be exact. The third is that there is a clearly marked boundary between science and ordinary thought. As these will have to be mentioned many times in this chapter I shall refer to them briefly as the certainty fallacy, the exactness fallacy, and the 'base degrees' fallacy.†

The first of these fallacies was sufficiently refuted by Hume. Scientific proof is not deductive proof, because every law is based on a finite number of trials, and there is no deductive way of inferring to any other trials, much less to an infinite number. There have been many attempts to state some principle that would turn scientific proof into deductive proof, but none comes anywhere near doing what is required of it. The second fallacy ignores observational errors, the treatment of which is an essential part of the technique of the subjects where greatest accuracy is attained. Most of this book has been devoted to showing how these two fallacies are avoided by the theory of probability.

†   But 'tis a common proof
That lowliness is young ambition's ladder,
Whereto the climber-upward turns his face:
But when he has attained the topmost round
He then unto the ladder turns his back,
Looks in the clouds, scorning the base degrees
By which he did ascend.

Julius Caesar, Act II, Sc. I.

An instance of the third is the rejection of colour under the description 'secondary quality'. Some people are colour-blind. Below a certain intensity of illumination (something less than full moonlight for me) there is no sensation of colour. A green object seen through red glass looks black, a scarlet pillar box seen in sodium light looks slate colour, and so on. It is therefore argued that colour is deceptive. However, an alternative view is possible. If I have a particular sensation of colour, I have it, and there is no more to be said. If, however, I say 'a brick is red', the word *red* does not mean the same as it does when I say 'I have a sensation of red'. The latter expresses direct knowledge.

'A brick is red' expresses an inference about an object, and the inference may be wrong, and in any case needs further interpretation. Whatever quality the brick may have that corresponds to my sensation of red, it is not identical with the quality of the sensation. A physicist might say that the quality is that it predominantly reflects light with wave-lengths about 6500×10^{-8} cm. Now it might be held that the latter statement is not subject to the ambiguities of the simple notion of colour in an object. It will be true that if the intensity of the incident light is measured for each wave-length, and the same is done for the light reflected from the brick, in any illumination and conditions of observation the light from the brick will be relatively stronger in the region of 6500×10^{-8} cm. wave-length. This replaces the statement about colour by one about measurable quantities. It is suggested that we can use this and forget that we ever had sensations of colour.

Can we? If we do, an analytical chemist or a field botanist can no longer use his colour sensations. He must measure the spectra of precipitates or flowers with a spectrophotometer. But he knows quite well that colour sensation is good enough for his purpose. He also knows better than to examine objects in sodium light or through coloured glass when judging their colour, and the ambiguities do not arise. The ambiguities all concern losses of discrimination between colours, and if the conditions are such that colours ordinarily discernible are in fact discernible the conditions are satisfactory. Essentially there is a cross-classification; observing conditions and colours are both classified.

Now this applies also to measurement. For refined measures of wave-length it is necessary to pay careful regard to the conditions of observation. The apparatus must not be under severe stress, otherwise lengths will alter. Systematic errors such as errors in graduation of the circle and variations of thermal expansion have to be watched and taken into account. Even when corrections for these have been made the measures still vary. Thus, in a sense, measures also are deceptive. In fact all the alleged deceptive features of colour have analogues in measurement of wave-length. The difference is one of degree, not of principle. The results of measurement of wave-length and of distribution of intensity along the spectrum are additional information to those given by observation of colour; they give no ground for dispensing with it.

I have referred to Eddington's treatment of relativity as exemplifying this fallacy; the possibility of treatment on the lines that he adopts is suggested by results of more elementary subjects, but at the outset he treats the indications of these subjects as conventions. Another instance is found in the common remark 'the laws of physics are really disguised definitions'. Definitions of what? If there are n parameters in a law, n sets of values of the variables suffice to estimate them. The information that the law fits for a much larger set of values is ignored.

An extreme case was Eddington's discussion of 'time's arrow'. The ultimate equations of physics are symmetrical with regard to past and future. What meaning can be attached to the increase of time? The fact that we are aware of it directly is rejected as irrelevant to physics. A physical substitute must therefore be sought, and is found in increase of entropy. Consequently, before we can say that one event is later than another, we must have measures of the entropy in the Universe at the two times.

9·1. *Realism and idealism.* At this point it is worth while to examine the notion of philosophical reality to see how far it is relevant to science. All our empirical data are sensations. The test of a scientific law is its capacity to account for sensations already recorded, and its interest lies largely in its capacity to predict new ones. It is a fact that over substantial intervals of time my sensations do not noticeably vary, but if they had never varied it would, I suppose, never have occurred to me to ask whether they could vary. The laws of science, stated in probability language, are statements about the probabilities of sensations in specified circumstances, and the parameters occurring in them need have no other status. However, we give them names: the mass of Jupiter, the Earth's magnetic moment, the atomic mass of iron, and so on. Have such names any meaning *beyond* their occurrence in probability laws? This question is closely related to the ancient question: does an external world exist? Some idealist schools maintain that it does not, but that the external world is a figment of human imagination. The extreme form of idealism known as solipsism would maintain that even other minds do not exist. A solipsist could say that his external world was only in his own imagination, and could not even consider the possibility that other people could imagine external worlds. I know of no professed solipsists. Possibly children up to the age of a few months are, but they do not profess anything. Realism† maintains that an external world does exist, and that the parameters in scientific laws are (mostly) expressions of properties of that world, and that in finding out something about their values we are also saying something about the world.

Neither realism nor idealism is refutable, though libraries have been filled with bad arguments on both sides. Apparently realism contains a hypothesis not contained in idealism, but when we try to state this hypothesis in terms that would permit a scientific test it always eludes us. For this reason Carnap calls the problem a pseudo-problem. I am inclined myself to think that there is meaning in it, though I cannot say what the meaning is. Most people, including most scientists, consider themselves to be realists, and realism has the advantage that it has a much more fully developed language. An adequate idealist language could be created, but the fact is that present idealist language is utterly inadequate for what we need to say.

Bertrand Russell‡ has pointed out that rejection of unproved hypotheses, if carried out thoroughly, would lead to the rejection of far more than had previously been supposed. The external world goes, of course, and so do other minds. But I cannot prove that I did not come into existence five minutes ago, being created with all the memories that I believe I had. All my information now would be precisely the same. In other words, if I made this drastic

† Mysteriously called Victorian by H. Dingle, *The Sources of Eddington's Philosophy* (1955).
‡ *Human Knowledge* (1948), p. 194.

rejection, I should be left believing even less than would have been thought possible for a solipsist. The point is that nobody is willing to accept the conclusion, and if anybody draws the line at some intermediate step, as professed idealists do, he can fairly be asked why he draws it there rather than at some other point.

What must be rejected outright are the naïve realism and naïve idealism that would maintain that inferences about sensations, or about the parameters that occur in the laws relating them, can be made with certainty. The possibility of revision is an essential part of scientific method. It may appear that no scientist could believe anything so silly as naïve realism or idealism, but the discussions of the general theory of relativity and the uncertainty principle showed that many did.†

One peculiar mixture of idealism and realism needs special mention: it is the statement 'things do not exist except when they are observed'. This is a characteristic confusion arising from the inadequacy of idealist language. To have a sensation of colour means the same thing as to observe a coloured patch. The *thing* whose surface is the coloured patch is *inferred* on a realist basis.‡ But the essence of a scientific law is that it contains variables; its predictions are for any possible values of these variables, at least within a certain range. In stating the existence of the thing at all we are contemplating the possibility of making an observation at any time. The time of the actual observation is in no way special in the statement of the law. To take a specific instance, to account for the motion of a planet it is necessary to allow for perturbations by the other planets at all intermediate times. The positions of the other planets are calculated in the same process, but the number of instants of observation is finite. Thus the process involves the supposition that the other planets have a gravitational effect at times when they are not observed and therefore, according to the above statement, do not exist.

We might try to avoid the difficulty by saying that observations of the position of Saturn, say, could be used to infer the perturba-

† Professor Dingle, with much of whose work I agree, is an idealist; but he seems to me to confuse realism in my sense, which I might call critical realism, with naïve realism, and consequently most of his arguments, while valid against naïve realism, fail to establish any case against realism.

‡ The word *perception*, as used by philosophers, appears to muddle up both sensation and inference.

tions produced by Jupiter and hence the positions of Jupiter at intermediate times. Thus the gravitational effects of Jupiter on Saturn might themselves be regarded as observations of Jupiter. But, since they are supposed to go on all the time, according to the method used in computation, this escape would imply either that Jupiter is observed all the time, and therefore that it exists all the time, or that the method of computation is fallacious, which nobody has dared to suggest.

The standpoint that things do not exist except when they are observed has the consequence that things never observed at all do not exist at all. But we have seen that that principle needs modification to admit quantities that are calculable at times when there is no direct measurement, and a corresponding modification in the conclusion leaves only the principle that we are concerned only with quantities that we can calculate, and not much is left of the drastic principle of the 'rejection of unobservables' that is so often claimed to be a fundamental principle in modern physics. The theories that appeal to it are in fact full of unobservable quantities (the coordinates themselves in relativity theory, arguments of complex wave functions in quantum theory) which have to be eliminated before observable results are predicted.

The point may be illustrated further by the discoveries of Neptune and the companion of Sirius. These were led up to by perturbations of, respectively, Uranus and Sirius. The suggestion that the disturbances might be due to the attraction of unknown bodies led to calculation of the positions of these bodies and to a search, which revealed the disturbers visually. But suppose that the disturbers had been too faint to be seen? (This is actually the case for several known stars.) The problem of predicting the position of Uranus would have remained, and would have had to be treated either by including unexplained periodic terms or by using the calculated orbit and mass of Neptune as data. The latter would be the more satisfactory because all the terms would be calculated in a single process, which would make full use of the theoretical relationships that exist between them.

9·11. *Phenomenalism.* This theory of knowledge is a form of idealism and may be defined by the statement that nothing is to be supposed to exist that cannot be reduced to descriptions of sensations. It may be traced back to the medieval writer William of Ockham. The rule known as Ockham's razor was apparently given first in the usual Latin form *Entia non sunt multiplicanda sine necessitate* by John Ponce of Cork in 1639. Ockham (d. 1349?) and a number of contemporaries made equivalent statements, but not in these words.† Its use can be illustrated by the idea of absolute position. In pre-relativity mensuration and dynamics this appeared in the laws, but all the verifiable inferences were in the form of measured distances and directions, involving only differences of the rectangular co-ordinates with reference to the supposed absolute origin. Consequently the predictions would be the same for any other laws that left the differences of the co-ordinates unaltered, and their verification, therefore, by the principle of inverse probability, leaves the posterior probabilities in the ratios of the prior probabilities. Such laws contain parameters that we cannot determine, stating the position of the absolute origin, and the actual co-ordinates that appear in them are indeterminate. If we perform an elimination so as to obtain only relations between measured distances between identifiable marks, we make the relation between facts and laws more immediate. A law stated in such a form is the disjunction of all those involving absolute position, and its probability is the sum of those of all such laws stated for different absolute positions. Thus the elimination of indeterminates from the laws has two advantages: it exhibits their verifiable content more clearly, and it increases their probability.‡ We may notice, however, that this gain is attained at a price. In celestial mechanics, if we have n bodies and therefore $3n$ co-ordinates, the Newton-Lagrange method gives all the equations of motion independently and symmetrically. If we work entirely in terms of differences we effectively refer everything to the centroid of the system, and therefore have three relations between the co-ordinates, which will introduce three undetermined multipliers instead if we try to preserve symmetry between the treatments of the various bodies. In practice this is not done: the variables are taken to be the co-ordinates relative to the most massive body, and thus symmetry is abandoned.

† W. M. Thorburn, *Mind*, **27**, 1918, 345–53.
‡ This argument was given by Dr Wrinch in a colloquium about 1917; I do not know whether it was published.

The modern form of phenomenalism is effectively due to Ernst Mach and Karl Pearson. Having myself started from the phenomenalist position, I must express my great indebtedness to these writers, but I consider that pure phenomenalism is not adequate for scientific needs. It requires development, and in some cases modification, before it can deal with the problems of inference; and Mach's failure to distinguish clearly between actual and inferred experience is one of the chief modern sources of confusion. The good points of phenomenalism are: (1) it requires analysis of suggested laws to exhibit what they actually say about experience, and (2) if a law contains reference to quantities whose values do not affect the prediction of experience, phenomenalism says that the law should be restated in such a way that these quantities do not appear in it. The bad point is an overstatement of (2) to the effect that everything mentioned in a law must be separately observable. An electron, for instance, is a valid scientific concept. I think that it is merely playing with words to say that it is a class of sensations, or that it can be described in terms of sensations. The point is that the concept permits co-ordination of a large number of sensations that cannot be achieved so compactly in any other way. The same applies to the matter at the centre of the Earth.

Speaking roughly, a hypothesis that explains more than one fact is useful. One that explains no facts is redundant. One that explains just as much as it assumes is *ad hoc* and tells us no more than the fact itself does. This was the situation about the aether in relation to electromagnetic waves.

It is sometimes said that scientific hypotheses should be actual logical constructs from observations: that is, that they should be built up from observations by use of the logical notions of *or*, *and*, membership of a class, *all*, *some*, and so on. This seems to have been definitely disproved by Braithwaite, who produces a type of admissible scientific hypothesis that cannot be built up in this way.†

9·12. *The theories of Russell and Whitehead.* Bertrand Russell, in *Mysticism and Logic* (1917), tries to tackle the problem of actually defining objects in terms, not exactly of sensations, but of sense data, which are effectively sensations with the errors of observation

† *Scientific Explanation* (1953), pp. 62–8.

removed. A physical object still cannot be adequately defined as the class of those sensations that, in ordinary language, would be said to be perceptions of it; for then the object would change when new aspects of it were observed, and this is not to be allowed. Therefore he considers the object defined in terms of all possible aspects of it; these aspects are called *sensibilia*, and resemble sense data in everything except that the majority of them are not perceived. An object is then a class of sensibilia.

From the practical scientific standpoint the weakness of this attitude is that we do not know what the sensibilia are like. An object, on this theory, could never be described until we had knowledge, by experience, of all its aspects, perceived and unperceived, and this is inherently contradictory. Even the perceived sensibilia, or sense data, cannot be described in terms of sensations until we have some rule for removing the errors of observation. The unperceived ones are necessarily never known directly, but have to be inferred from the perceived ones; and this can be done only by using the laws of physics, inferring properties of the object, and then proceeding to the unperceived sensibilia. The physical object and the laws of physics are anterior in knowledge to the sensibilia, and Russell's theory, whether it is logically consistent or not, is not a theory of scientific knowledge.

I should say that in *Human Knowledge* Russell does not mention sense data and has presumably abandoned this theory. However, the theory and the objection to it are interesting because they bring out the point that in a theory of knowledge we *must* proceed from the better known to the less known, and a theory that reverses this process is not a theory of knowledge.

In Professor Whitehead's theory† events, instead of sensibilia, are the fundamental entities. Each event contains other events, so that we can specify sequences of events such that each event in a sequence contains all after it. The limit of such a sequence is a point-event, and it is to such point-events that the laws of physics are supposed to apply. But the notion of a limit implies an infinite sequence, and an infinite class of observations is impossible in practice.

† *An Enquiry into the Principles of Natural Knowledge* (1919).

9·2. *Infinite sets.* We may say that it is never possible to construct a valid theory of knowledge that involves the use of an infinite set of observational data. The objection is similar to that given by Poincaré in his criticism of Cantor's theory of infinite numbers. † Poincaré argued that it is impossible to assert anything about a class, and in particular anything about the number of its members, until every member of the class has been defined in words; and, as only a finite number of entities can ever be defined in words, it is impossible to know anything about an infinite class, and in particular anything about infinite numbers. The argument as it stands is not valid against Cantor's theory, because in order to make an assertion about a class it may not be necessary to have a separate definition of every member; often a general proposition about all members can be asserted or postulated, and is enough for the purpose.. Poincaré, indeed, seems to have overlooked the fact that if his argument was sound it would also destroy the whole theory of infinite series and of differentiation and integration; thus little would be left of higher pure mathematics. Thus the convergence of a series depends on the proposition that if we take any fixed positive quantity ϵ, the sums of the first n, $n+1$, $n+2$, ..., terms, for some value of n, depending on ϵ, all differ from a certain number, called the sum of the series, by less than ϵ. These sums are infinite in number, and if separate definitions were needed we could never prove that a series is convergent. This result is of course quite unacceptable. But the argument would go even further than this. Nobody has had time in his life to construct definitions of every member of a class of a million members; and as a number is merely a property of a class, it should be impossible to prove that, for instance,

$$1\ 000\ 001^2 = 1\ 000\ 002\ 000\ 001.$$

Thus the argument would also invalidate most of arithmetic. If therefore we believe that the propositions of arithmetic have some meaning and are true, we cannot accept Poincaré's objection to the theory of infinite numbers.

But while the argument is wrong in this case, it is clearly valid when our only source of information about the members of the class is empirical; for the whole number of observations that a person can make in his life is finite, and hence his experience alone

† *Science et Méthode* (1908), 192–214.

will never tell him anything about all the members of an infinite class of entities. Any proposition about such a class, or about all its members, is necessarily either wholly *a priori* or else an inductive generalization, and in the latter case it is neither known directly nor obtainable from experience by the principles of deductive logic alone. The data of any branch of science must consist of a finite number of observational results and some *a priori* principles.

One consequence of this is that we can never prove by experience the existence of a limit to which a series of entities known by experience may tend; for in order to establish the existence of such a limit we should need to have knowledge that an infinite number of such entities are within any arbitrarily small distance of the limit. This by itself would not be a fatal objection to any such theory, for there seems to be no possibility of constructing a theory of knowledge without some assumptions, and it may be considered that in the case in question some conditions are satisfied under which the existence of the limit is known *a priori*. But what is fatal is that in actual problems we do not only want to know that the limit exists; we want its value according to some definite system of measurement, and that value can never be known *a priori*; indeed, if it was, there would be no need to make measurements at all. Thus if a limit is used in scientific theory, its value and all propositions that are about it are neither wholly *a priori* nor known by experience, and therefore are not primitive propositions that can be used in a theory of knowledge based on experience. It will be seen that this rules out at once the theories of Russell and Whitehead considered in 9·12.

9·21. *The frequency theories of probability.* The main point of the present work is that to give any systematic account of scientific method it is necessary to have a notion of partial proof, which we call probability, and is essentially an *a priori* notion not belonging to deductive logic. This notion is habitually neglected, and statisticians, while sometimes using the word *probability*, attempt to give definitions of it that avoid introducing any idea that does not belong to deductive logic or to classification of observations.

The attempt most often mentioned is the limit theory. This seems to have been advocated first by Leslie Ellis and Cournot, but was developed by John Venn,† and later by von Mises and Reichenbach.

† *Logic of Chance,* pp. 162 *et seqq.*

Venn considered that the notion of probability presupposes a series, the terms of which are infinitely numerous and represent the cases of an attribute ϕ. From these we can pick out a sample of m members, of which l, say, possess the further attribute ψ. Then the sampling ratio is l/m. Then m is imagined to increase indefinitely, and the probability of ψ given ϕ is defined as the limit of l/m as m tends to infinity.

The form of this definition restricts the field of probability very seriously. It makes it impossible to speak of probability unless there is a possibility of an infinite series of trials. When we speak of the probability that the Solar System was formed by the close approach of two stars, or that the stellar Universe is symmetrical, the idea of even one repetition is out of the question; but it is in just these cases that the epistemological problem is most acute. But this is not all, for the definition has no practical application whatever. Advocates of such a definition really mean what they say; they say again and again that the notion of probability applies only to infinite series, and has nothing to do with individual cases. This is obvious on this definition. If the limiting frequency in an infinite number of trials is 0·9, that is perfectly consistent with ψ never appearing at all in the first 100 trials and maintaining an average of 90 per cent. afterwards. But practical men such as farmers and engineers come to statisticians for advice, and in every case they are concerned with a finite number of instances. When an applicant for insurance wants to choose between a policy that offers a large return if he retires at age 65, as against one that offers return of premiums if he retires before 65 and a smaller pension if he retires at 65, *his* probability of living to 65 is an important consideration. The limiting frequency in an infinite series of people is of no direct interest to him. The insurance company itself is concerned with a large number of cases, but up to a given time even this number is finite, and therefore the probabilities are meaningless according to the definition.

Again, what reason is there to suppose that the limit exists? It postulates an infinite number of random choices, and if the multiplicative axiom is not assumed this is impossible. Even if the axiom is accepted, then there is still a choice at every stage and there can be no deductive proof that the ratio tends to a limit, as we saw in 3·1. What does appear to be true is that there is probability 1 that the sequence will tend to a limit. But this depends on (1) the use of probability as a primitive notion, (2) the axiomatic rules of addition and multiplication, (3) the notion of randomness, subject to the existence of a chance. On the limit theory there is no justification of any of these. The existence of the limit is in fact an *a priori* assertion about the result of an experiment that nobody has tried, or ever will try.

The difficulties become worse when we attempt to combine probabilities, for then we have to face an indefinite repetition of infinite series. This was called by Venn the use of cross-series, and forms an important part of his theory of inference. It is necessary, for instance, in giving a meaning to the proposition connecting the probabilities of a proposition referred to different data,

$$P(p.q|r) = P(p|r)\,P(q|p.r).$$

For an infinite series is necessary to give an account of $P(p|r)$. From the members that satisfy p we have to select the members that satisfy $p.q$. Thus we have to examine in one direction to find the frequency of q given p and r, and in the other direction to find those of $p.q$ and p given r. Thus a single series is not enough; we need a doubly infinite series. There is no more reason to believe in the existence of limits in this case than there was in the other; and the opinion that the limits, if they exist, will satisfy the relation depends on the ordering and can be justified only if the samples are made according to some special rule. Thus the statistical theory of probability becomes a network of begged questions.

Fisher† seems to have felt the difficulties about infinite series, but having misunderstood the theorem of inverse probability he took the drastic step of saying that a probability presupposes a hypothetical infinite population, and the probability is a sampling ratio in this population. This of course is a leap from the frying pan into the fire, because ordinary notions of ratios cannot be applied to infinite numbers at all.

It may be asked how it is, if these methods cannot be carried out, that their exponents manage to get any answers. The reason is that they never use the methods. They take over the addition and product rules from the epistemological theory, and naturally get

† *Phil. Trans.* A, **222**, 1922, 312.

the same answers. They differ from me in interpretation; I say that the answers have meaning for a finite sample, whereas they say that they have no meaning for a finite sample and therefore for any actual application.

The third method, often called the classical one, is a development of that of de Moivre, who treated probability as simply a ratio of cases of ψ given ϕ to the whole number of ϕ. Thus the probability of throwing a head with a coin is $\frac{1}{2}$ by definition, so long as the bias is not so extreme as to make one face impossible. It reduces probability theory to a part of the theory of permutations and combinations. The modern extension regards probability as a ratio of measures of sets and derives the addition and product rules as consequences. This is quite satisfactory mathematically, but leaves the subject matter completely undescribed: it does not explain what the theory is about.

The point of all these systems is that the common-sense view of probability is vague; but instead of recognizing that it is indispensable and trying to make it more precise, they reject it altogether. This is the worst case of the 'base degrees' fallacy, since the substitutes are such as could not work in any circumstances whatever. They refuse to speak of the probability of a hypothesis, but they substitute 'limits of acceptance'. Thus for the χ^2 rule a value χ_0 is chosen such that there is a chance, subject to the conditions for the rule, of 0·05 or 0·01 of χ^2 exceeding χ_0^2. If χ^2 exceeds χ_0^2, the hypothesis is rejected; otherwise it is accepted. The limits 0·01 and 0·05 are, however, totally arbitrary; and there is no explanation of what, if anything, is being said about the hypothesis when it is rejected. Since the notions of reasonable belief and of inference to new data are rejected as meaningless, there is no apparent reason why rejection or acceptance should make any difference to the future use of the hypothesis.

No differences of prior probability that have ever been seriously suggested make more difference to the results than one or two observations more or less do, apart from the distinction between estimation problems and significance tests. The latter is an essential matter of principle. On the other hand the principle of inverse probability says that all the information provided by the observations is summed up in the likelihood; any alternative summary is at best an approximation to that given by the likelihood, and at worst is no better than a guess. But all the systems that I am criticizing, except Fisher's, pay no special attention to the likelihood, and the result is a loss of accuracy, more or less equivalent to sacrificing a fixed proportion of the observations, however many they may be. Further, though they use the expressions 'estimation problems' and 'significance tests', none states clearly what the difference is, and in practice they use the same rule for both. In fact the vagueness of the prior probability is negligible in comparison with what is tolerated in these systems.

Carnap† distinguishes between two kinds of probability, of which the first is similar to the logical relation treated in this book. It is formalized in detail, I think somewhat prematurely, since he gets no further than Laplace's theory of sampling. The second kind is defined by the notion of limiting frequency. The arguments given above seem to dispose of this definition. There may, however, be meaning in the distinction, and an intrinsic probability of events may exist. The chance of throwing a head with a coin is apparently a property of the coin and of the conditions of throwing; attempts to define it lead nowhere, but it is intelligible in its own right. The situation is that if the chance of a head, in this sense, is $\frac{1}{2}$, then on this information the probability, in the logical sense, is also $\frac{1}{2}$, and hence intrinsic probability can be discussed in the language of epistemological probability.

9·3. *The Philosophy of 'Nothing but'.* I use this expression for several forms of the 'base degrees' fallacy that are based on an exaggerated phenomenalism. They all appeal to the principle that nothing is relevant to science but what is observed by physical methods. Since we cannot observe consciousness physically, consciousness must be rejected, and the result is behaviourist psychology. A place can be found for thought, because when a person thinks he is thinking there are small movements in his throat that can be detected instrumentally.

Some logicians of the logical positivist school regard logic as only the manipulation of symbols; even the notion of entailment disappears, being replaced by a 'rule of procedure'. This means that we can carry out deductive processes without knowing what deduc-

† *Logical Foundations of Probability* (1950).

tion means. But if somebody makes a mistake, what is the remedy? To point out that the rules have been wrongly applied; that is, the rules say that some procedures are permissible and others are not, and in any particular case it is necessary to decide which. But this is a deduction from the axioms and the decision to adopt these rules of procedure. Deduction thus retains its meaning and is applied, though it is not formally mentioned in the rules.

It is necessary to distinguish entailment from implication. In ordinary usage they are the same thing. (p entails q) means 'q can be deduced from p'. But in *Principia Mathematica* 'p implies q' means '$\sim p$ or q'. This would hold if we have independent knowledge of either $\sim p$ or q, without there being any relation of deducibility connecting p and q. The distinction was, I think, first made clear by G. E. Moore. (p entails q) has nothing to do with whether p is true. But if $\sim p$, whether q is true or not, then ($\sim p$ or q), so that we reach the peculiar proposition that a false proposition implies any proposition. This, in spite of the superficial resemblance, is quite different in meaning from the proposition that a contradiction entails any proposition.

Deduction cannot be completely formalized, as was shown by Lewis Carroll† in 'What the Tortoise said to Achilles'. p and (p entails q) together entail q; but if we formalize this we have simply

$$(p \text{ and } (p \text{ entails } q)) \text{ entails } q. \tag{1}$$

What we want to say is

$$\text{'}p \text{ is true and } p \text{ entails } q; \text{ therefore } q\text{'} \tag{2}$$

and then to detach q and assert it by itself. Suppose a sceptic asks for reasons why he must do this. We state the principle that we can; this is a longer proposition, which still has not detached q, and the sceptic can raise the same difficulty again. The position is that we can *see* that the process is valid, but attempting to formalize it results only in building up longer and longer sentences, which never do what we want. It is for this reason that in logic deduction is stated as a rule of procedure and not as an axiom. But this device tends to obscure the essential nature of entailment.

The assertion of a proposition in fact can never be formalized.

† *Complete Works*, pp. 1225–30; *Mind*, **4**, 1895, 278–80.

We can denote a proposition by p without stating whether it is true or not. We can have a rule of procedure to put the mark $\vdash$ before p if p is asserted to be true. But in such cases p is either known directly or is the result of a deduction. In either case p is seen to be true, but that is all; trying to provide an axiom to state that we can apply the sign $\vdash$ leads straight to the Tortoise paradox. Thus even deductive logic needs a notion of appreciation that cannot be formally expressed within the system, but depends on looking at it from outside; it belongs to the meta-logic. Rather curiously, the only work on mathematical logic that I have seen since *Principia Mathematica* that mentions the point is H. de Long, *A Profile of Mathematical Logic*, 1970. The $\vdash$ sign is introduced on p. 108. He agrees that it belongs to the meta-logic.

Any logical system is imbedded in a language; that is, any statement of the system is expressed by a sentence in the language. This applies not only to systems of pure logic but to any branch of science; classical mechanics, for instance, is a precisely stated system and is written about in a special language. For technical reasons it has been found convenient to speak of the rules of the language and to manipulate according to them, and there is a tendency to identify the language, the logical system, and the subject matter. Carnap, for instance, has said that 'The lecture was about Babylon' is more clearly expressed by 'The word *Babylon* occurred in the lecture'. I have no intention of denying the achievements of him and his colleagues, and his emphasis on the distinction between a thing and its name, or between a proposition and the sentence that expresses it, is most valuable, though it would ruin all crossword puzzles. However, a warning of possible dangers is needed.

Suppose that we have a logical system S and that the propositions in S and deducible from S are expressed by sentences in a language L. Suppose that L contains the word *not*. Then if S and its consequences contain the proposition expressed by p, not-p is a properly constructed sentence in L. Therefore, if L contained no sentences other than those expressing propositions in S and their consequences, not-p would be deducible from S and S would be self-contradictory. It is therefore not adequate to identify the logical system with the language used to express it, or even to suppose that every sentence in the language expresses a proposition in the logical system. The point is simple, but I have several times heard remarks in discussions that seem to indicate that it is not generally appreciated.† The language must contain sentences that do not express propositions of the system. This enables us to consider statements before we know whether they are true or not.

9·31. A prevalent statement about probability theory, analogous to the attempt to banish entailment from deductive logic, is that probabilities cannot be admitted into science unless a means of measuring them is provided. In relation to intrinsic (often called objective) probabilities this leads to the frequency definitions, and as pointed out above these are inapplicable in practice. But the mistaken belief that they constitute a practical method of measurement leads to the demand that epistemological probabilities also must be measured, and then their identification with actual degrees of belief leads to the further demand that actual degrees of belief must be measured before the theory can be considered. This completely misses the point of the theory. The theory is a way of dealing with the observational data of science; it is not itself based on observation and could not be. Probabilities can be calculated, just like any other numbers. Without rules for analysis of the data science would be replaced by a mere collection of unsystematic observations. If people's actual degrees of belief could be measured, the measures would be data of psychology and rules for analysing them would still be needed.

The alternative to having a theory of inference, while maintaining some sort of system in science, would be to regard the rules for classifying data as conventions, and inference as a peculiar habit following no rules. This is perhaps most explicitly maintained by Eddington on the one hand and orthodox statisticians on the other. The conventions are apparently arbitrary. The theory of probability, on the other hand, explains why the rules for analysing the data are suitable and also gives warnings against dangerous exaggerations.

9·32. It is often said that a proposition is meaningless unless a method can be provided for determining with certainty whether it is true or false. If that was so, every statement about the

† Carnap says that he proves his Language II consistent, and this statement is misleading for the above reason. What he proves is that the *syntax* of Language II is consistent, the syntax corresponding to the logical system.

external world would be meaningless. Some quantum physicists say that a probability distribution for a measure is meaningless unless the measurement can in some circumstances be made exactly. There is no such measurement. But these writers say that any two quantities except dynamical conjugates can be measured simultaneously, which is proposterous if exact measurement is the criterion. Such rules manage to combine the three fallacies of exactness, certainty, and 'base degrees'.

9·4. *Determinism.* This is related to what philosophers call the Principle of Causality, or the Uniformity of Nature. It may be expressed in the form: given the state of the world at any instant, the state at any later instant is determinate. This requires some discussion of the meaning of *state*. The positions of all particles in the world at some instant obviously do not determine the motion afterwards unless we also know the velocities. But particles are not enough. The time of an event has to be distinguished from the time when it is observed. If a lamp is extinguished, it is still seen for some time at distant places; the sensations produced are the same as if the light was still shining. If we consider the state at some intermediate time, we must say that the illumination is caused by light *on the way*, for the lamp is no longer available as an explanation. The state to be specified as determining the future must therefore include the positions and directions of motion of light waves. The alternative is to say that the state of a system at any instant is determined, not by the state at each single previous instant, but by the aggregate of states at all previous instants. The position is tenable; but now we see that the previous instants to be considered stretch right up to the instant of observation, and we may then reasonably say that the state then is determined by the states at instants indefinitely shortly before. But then the notion of light on the way is unavoidable, and we may as well say at once that the law of causality is expressed by differential equations with regard to the time. If we insist on specifying the state only in terms of material particles we must consider laws as involving finite intervals of distance and time explicitly, and are committed to the notion of action at a distance.

The principle of causality now becomes the aggregate of all scientific laws, whether already known or awaiting discovery. To accept it implies a hope that we may some day know all laws; but that day is still distant. As a working rule it may be valuable for its psychological effect, but science needs the principles of inference anyhow, and the principle of causality adds nothing useful. It appears indeed that quantum phenomena require an irreducible element of randomness, and that strict causality will always elude us.

The words *cause*, *effect* and *because* are on a different footing, and have nothing to do with a general principle of causality. If a scientific law involves a certain number of variables, then a knowledge of all but one of them within stated limits determines that one within a corresponding range. We say that it has a particular value *because* the others have specified values. The notions of cause and effect involve rather more than this; there is an asymmetry about them that is absent from the word *because*. Thus we may say either that a triangle has the angles at the base equal because it is isosceles, or that it is isosceles because the angles at the base are equal. When we speak of a cause and an effect, we pick out one as the cause and the other as an effect, and they cannot be interchanged. The distinction seems to be one of time; the events under discussion are connected by a law, and we call the earlier the cause and the later the effect. There is no distinction for contemporaneous or overlapping events.

Causality is often associated with some notion of inherent necessity. It seems to be impossible to say just what this means, except perhaps in terms of some sort of animism: some sort of demon may inhabit a tree and the growth of the tree might be in response to the demon's wishes. But the idea is scientifically useless. If we can say with high probability that a set of circumstances would be followed by another set, that is enough for our purposes. The demon is an *ad hoc* hypothesis and nothing can be inferred about it except by studying the tree—and then it is easier to say that we are studying the tree. But perhaps the idea by itself does no great harm. The danger arises from the converse statement 'every event must have a cause'. If this is accompanied by an unsatisfactory explanation of the event, we may point out that the explanation does not explain, or that it postulates something that we have good reason to believe false. But our objection may be met by 'Then how do you explain it?' The answer is that science at any moment does not claim to have explanations of everything; and acceptance of an inadequate explanation discourages search for a good one. Causality is sometimes said to mean invariable succession and nothing else. Even in this form it is useless, since we do not know of any particular succession that it will always be invariable, and we need also to deal with successions that have not been so.

9·5. *Repetition.* It is sometimes said that the test of a hypothesis is in making a new experiment with all the variables the same as before, and examining whether the results are the same. Now *all* the variables include position and time; thus the new experiment would be identical with the old experiment and tautologically must give the same result. This is presumably not what is intended; the new experiment is intended to be distinct from the old one and thus if the observer is the same it must differ in time. But in practice many other things besides time will have changed, such as temperature and atmospheric pressure within the laboratory, and there will have been many new events both inside and outside the laboratory. Also the result of the experiment will not be exactly the same. If we make many repetitions and have no detailed record of the other circumstances, no hypothesis other than that of random variation is available for consideration, and the only thing we can estimate is the random variation in the data. Such a result corresponds to the situation in what agriculturists call uniformity trials, in which they want to know the general magnitude of the variations of fertility between plots; these are given similar treatments and similar crops are grown on them, the object being to get some idea of the number of plots that will be needed in design of future experiments to find results of the accuracy desired.

A less naïve view is that we should vary the conditions one at a time (this is what happens in dynamics when the time itself is the independent variable). This works up to a point. But suppose that z is of the form $ax + by + cxy$, and that the standard of reference for x and y is $(0, 0)$. Then we can consider only pairs of the form $(x, 0)$ and $(0, y)$; and however many observations we have cxy will always be 0 and we shall never be able to estimate c. In general, the method will not succeed unless z is of the form $f(x) + g(y)$. If it is not, we *must* be allowed to change both independent variables simultaneously. Even then there seems to be no reason except convenience for repeating observations for the same pair of values. It is equally informative with regard to accuracy, and more informative with regard to the form of the law, to take many different pairs.

Repetition of the whole of the experiment, if undertaken at all, would usually be done either to test a suggested improvement in technique, to test the competence of the experimenter, or to increase accuracy by accumulation of data. On the whole, then, it does not seem that repetition, or the possibility of it, is of primary importance.

If we admitted that it was, astronomy would no longer be regarded as a science, since the planets have never even approximately repeated their positions since astronomy began.

There is a contrast between natural sciences and laboratory subjects, but it is not a matter of repetition. The laboratory worker has much more freedom to choose the scale of his apparatus and the actual values that he will adopt for his independent variables. We may say that his conditions are more under control; he is not limited to waiting for a natural phenomenon to happen, and he may need to consider fewer variables at once. But the consideration of several variables at once is merely more complicated; it involves no new principle. The control of the experiment is not a matter of principle; it is essential that, once the experiment is started, nature is left to take its course, and this is true whether the initial conditions are set by man or whether he has to do his best with what nature provides.

9·6. *Psychology.* **9·61.** *The beginnings of induction.* Inductive reasoning begins in infancy. A child has to learn associations between its experiences as an essential part of adapting itself to life. In particular it learns associations between sounds and objects, and hence acquires some understanding of language even before it can speak itself. This has two consequences. The first is that the process is so familiar that most people think that it needs no attention, and when attention is given to it they think that something else is intended. Most people go through life without recognizing that there is any difference between induction and deduction, and when it is pointed out that an argument is not deductive they say that it is 'common sense' and leave it at that. But in infancy the distinction has to emerge from experience; generalizations are made and many of them turn out to have exceptions. In psychoanalytic language the Ego is the part of the personality that deals with the relations between the individual and experience; and any failure of a generalization must have a shattering effect on the infantile Ego. Psychoanalysts mostly say that when the emotional causes of anxiety have been removed the adjustment to reality follows with little difficulty. I have great doubts; induction is more difficult than deduction and a deductive argument has to be fairly simple to be understood by most people. I am confirmed in my suspicion by the persistent attempts of people, otherwise well above average intelligence, to represent inductive arguments as deductive, and by the existence of two types of scientist, one of which is sure that some hypothesis is right, while the other maintains, apparently, complete scepticism about all hypotheses. The real point, I think, is that most repressed anxiety does arise from uncertainty with regard to emotional situations. The word 'danger', so frequent in psychoanalytic literature, implies both a possibility of injury and uncertainty as to whether it will actually take place. An injury that is certain to take place is not a danger, nor is one that has already happened. Consequently the uncovering of the memory of the danger gives relief of anxiety, since it is then recognized that the danger is past.

It is in scientific method that the problems of induction present themselves most clearly, since we are continually having to handle new sets of data. But the essential problem of generalization goes back to infancy. We can try to state it more clearly and to produce a more systematic solution than was possible at an earlier age; but there is no problem of principle in scientific method that has not had to be faced and partially solved by every individual before he can tackle the problems of everyday life. The inference that other people's bodies are associated with minds is as difficult as anything in modern physics.

9·62. *Testimony.* This does not require the inference of other minds. It is based first on the association of sounds with events, and on the fact that different people make similar sounds in relation to similar events. This is the first step in understanding speech, and by generalization enables us to understand people other than those the language was learnt from. Understanding writing and other languages follow by further generalization. Thus a solipsist could arrive at the understanding of testimony without needing to make the step of supposing other minds to exist, provided he is willing to accept induction. I do not know whether anybody adopts this position, but it should be attractive to extreme advocates of the principle of the rejection of unobservables.

9·63. *The doctrine of general consent.* It is argued that a fundamental principle of science is general acceptance. This may mean several different things. The ultimate data are sensations, which are essentially private. There can be no consent until there is at least sufficient progress for testimony to be accepted, and this needs some inductions. Testimony itself clearly cannot justify them. But the doctrine can be stated so as to concern inferences made after communication has become possible. But even then it cannot concern sensations; it might deal with the principles of inference and with the later inferences themselves. But even in this form it is obviously inapplicable in practice. We cannot ask everybody before we make a decision. Even the people qualified to judge on the particular question concerned are usually too many to approach; and the decision about who is qualified is hopelessly impossible. The most that an investigator can do is to present his data and his arguments; and it is reasonable for him to expect that other people will accept them if they are right. But the discoverer of a new result is always in a minority of one. The data and the arguments are the essential point; wider acceptance is a by-product. Every individual has to make his own decision. The criterion of general agreement is like the instruction 'do not go into the water until you know how to swim'.

Publicity is important in a different way. When studying another person's work we have to take his word for the observations; they have been made once for all. In cases of doubt the experiment may be repeated. But the more fully the data are given the better position the reader is in to assess the probability of the result. Argument, on the other hand, once published, is available for inspection by anybody at any time.

9·64. *Materialism.* Our primary data being sensations, it may be said that the aim of science is to account for sensations in terms of ultimate concepts and their properties. On the materialist theory these ultimate concepts are those of physics and no others. The physiological interpretation of psychology does not go so far as this, but states that psychological phenomena can or will be reduced to physiology. The experimental study of sensation has gone some way in the explanation of the transmission of stimuli to the brain, but little has been done, except by psychological methods, towards understanding what happens to them when they get there. The opinion that the amazing complexity of mental processes, including recognition of sensations, emotions, reasoning and volition, can be reduced to physiological processes, is hypothesis; it may be true, but it is at present pure unverified hypothesis.

9·65. *Psychoanalysis.* All the mental processes just mentioned have in common the fact that they involve, to varying degrees, conscious criticism. This is directly recognized and therefore is a fundamental concept. One way of studying it is to examine mental behaviour when it is removed as far as possible, and to see what differences arise. The absence of criticism is best realized in dreams and in the psychoanalytic situation, where the patient, as a regular matter of technique, says everything that comes into his mind without criticism. The results are not chaotic; they are found to arrange themselves according to definite rules. These differ from the rules of conscious criticism, the function of which is to study them; and they are found to be closely related to the forgotten experience of childhood and the pitiless logic, based on incomplete data, of the *enfant terrible* and the child at still earlier ages. The result is the discovery of a whole region of mental activity, with laws of its own, demanding new concepts to express them. The Unconscious is the general name for this region; for details reference must be made to the special literature of the subject.†

The results of psychoanalysis have been criticized on various grounds, which seem to me to involve points of principle of much wider application. One line of attack (not so common now as formerly) is simply to deny the facts as discovered, or the truth of the relations between them. This is merely a matter of refusal to investigate, and does not impress the analyst who is dealing with the material every day, or the patient who has been cured of various mental disorders, ranging from minor anxieties to phobias or disabling neuroses, by being enabled to understand his own mental processes better.

† See especially Freud, *The Ego and the Id.*

Another objection is that the data are not public. For one thing, the analysis is conducted in a room containing only the patient and the analyst, and if any other person was present the patient would stop talking. Reports have been published; and if it is objected that these are reports and the original observations were private, that is true of any data whatever. What is more to the point is that the analyst selects special points in the patient's remarks for comment, and consequently there might be ground for suspicion that the data are not a fair sample. I think the answer is that patterns recur more often than they do in ordinary conversation; this could be a matter for a significance test, but a very complicated one, and the numerical data are lacking. The crucial tests, however, are the patient's emotional appreciation when a disturbance is released and the recovery of memories from ages previously completely forgotten; these do not occur in ordinary conversation.

Some psychoanalysts speak in terms of a naive idea of causality, such as has been totally condemned in this book. This could, however, be corrected without much change in the main arguments.

It can be said that the psychological processes are really the expressions of physiological ones, and that the solution of the problems investigated must come ultimately from physiology. This may be true. But to use it as a basis of procedure is not legitimate, because it assumes from the start that there are no ultimate mental concepts, or, what is the same thing, it takes for granted that there *are* relations that determine the phenomena of conscious mental activity in terms of those of physiology before we know what they are. Instead of inferring the laws from the data, the invariable scientific procedure, it begins with unstated laws and treats the data as a ground for optimism about the future. The situation is the same as if an engineer in process of designing a bridge was told that he should not attend to experimental evidence about the strength of his materials because all phenomena of elasticity may some day be explained in terms of quantum theory. It may be so ; but he wants to get the bridge built.

A further consideration is that even if such a hypothesis is true we should still be under an obligation to check whether its consequences are true. That implies investigating mental phenomena, and providing explanations of the facts that psychology has already disclosed. The hypothesis saves no work even if it is correct.

9·7. *Life.* The question of ultimate concepts arises again in such biological questions as the materialistic interpretation of physiology and the physiological interpretation of psychology. Modern research has shown that many physiological processes satisfy quantitative laws like those of physics and chemistry, and in many cases that these processes can actually be interpreted in terms of physics and chemistry. Are we justified in inferring that all physiology is reducible to physics and chemistry? It must be remembered that when the question was formulated the atom was considered an ultimate reality; the result of modern developments in physics is that we are asking whether physiological processes can be explained in terms of ψ-functions. The alternative is that there is a non-physical concept, which we call life, and which may be ultimate. The problem of materialism is to explain life. Life as it stands is a valid scientific concept because it explains observed phenomena; a live animal has properties not shared by a dead one. That is not to say that it is an ultimate concept. There seem to me to be two relevant indications, pointing in opposite directions. The growth of green plants involves the interaction of carbon dioxide and water to produce sugar or starch and oxygen, a reaction requiring the absorption of energy, which the plant obtains from the Sun's radiation. Carbon dioxide and water are ordinarily stable in each other's presence; the plant must apparently have some directing ability, applying the solar energy in just such a way as to upset this stability. The same applies to the obscure organisms that derive their energy from chemical reactions without the presence of light, reactions that do not take place spontaneously, but only under the influence of the plant itself. On the other hand, if organisms have a directing power, of molecular fineness, as this would suggest, they might apply it to the sorting out of molecules according to their velocities. Then they could upset the second law of thermodynamics and provide for themselves all the available energy they need. This does not appear to happen; physiological processes in animals and plants do appear to follow the second law of thermodynamics. The hypothesis that life is not an ultimate concept remains untested.

9·8. The words 'subjective' and 'objective' have hardly been used in this book. This is deliberate; it is because they have been used in so many senses that they have become a source of confusion. In philosophy the object was held to be studied by the subject; so a primary meaning in a realist system would be that the external world is objective and the thinker the subject. But then, since other people are part of *his* external world, they would be objective, and everybody would have a different definition of objectivity. To avoid this the meanings have been modified. On the one hand the judgments of all people are classed together if they are coherent according to some criterion of general agreement, and then called objective, while differences between people are called subjective. On the other, the whole process of thought is called subjective, whether there is, or could be, agreement about it or not. The final stage was achieved by Eddington, who regarded the laws of physics as subjective and psychology as objective. In the circumstances the less the words are used the better.

Much the same applies to the word 'perception'. I distinguish between a sensation, which is immediate experience, the process of inference to an object, and the object itself. But when the word 'perception' is used I usually find it quite impossible to discover which of these is intended, or whether it covers more than one.

9·9. *Models.* A simple type of model is afforded by diagrams, as in Chapter VI. These serve principally to aid the memory of the meanings of the symbols used; so long as they do this the diagrams do not need to be accurate. The accuracy comes from the calculations, in which the diagram is not explicitly used. Three-dimensional models are useful aids in understanding the structure of crystals. Essentially the symbolic expression of a mathematical argument is a model; it both aids the memory and makes the argument communicable to a distant person. In all these cases there are striking differences between the model and what it is intended to imitate. The value of the model lies in correspondence, not identity, between relations.

Another type of model is illustrated by experiments on the resistance of solids to the motion of fluids. Here the equations of motion can be reduced to a non-dimensional form. If the linear dimensions are L, the velocity far away from the solid V, and the kinematic viscosity ν, then if V is much less than the velocity of sound the equations can be expressed, by a change of variables, in a form where the only parameter is LV/ν. The force on an immersed solid will be of the form $\rho L^2 V^2 f(LV/\nu)$. The equations of hydrodynamics are usually too difficult to solve in detail, but if the force is found experimentally for one body for different values of LV/ν, f is determined and the law can be applied at once to a body of the same shape and any size and a fluid of any density, always subject to the restriction on V. Here the model is actually used to find quantitative answers.

In another type of problem the distinction between a model and the original is very vague. Larmor invented an aether with gyroscopic coupling between its parts, and Maxwell's equations came out as a property corresponding to elastic waves in an ordinary solid. Was this a hypothesis about the aether or a model of it? I think it was the former. For one thing it was only thought about, not actually constructed. But if we take it as a model it illustrates another point about the use of a model. If it does not imitate the essential relations of the original, and its consequences are worked out and adapted to the original, they may be expected to give wrong answers. On the other hand, if it does imitate them, the mathematical solution is exactly the same for the model as for the original, and the model gives no help whatever. In the above hydrodynamical example the usefulness of the model arises from the fact that an actual fluid can be used to solve the differential equations for us—the mathematical solution does not need to be carried out.

CHAPTER X

STATISTICAL MECHANICS AND QUANTUM THEORY

Someone said, Do not tell me that this son of a villager from Tsou is expert in matters of ritual. When he went to the Grand Temple, he had to ask about everything. The Master hearing this said, Just so! Such is the ritual.　　　　　*Analects of Confucius* (WALEY'S translation)

10·0. In preceding chapters I have shown how mensuration, classical mechanics and large parts of relativity theory can be derived from experience and the principles that are needed in order that it should be possible to establish high probabilities for scientific laws. So far as these subjects are concerned there is little disagreement between physicists with regard to the laws themselves, which are accepted as in accordance with experience. The chief point of the present treatment is that it brings out what features in experience are actually relevant to the acceptance of the laws. In particular it shows that errors of observation have to be considered in the process of establishment of the laws: we can, if we like, say that there is a form of the law that expresses exact relations between true values, but the law that is verified is a modification of this that takes account of probabilities of errors of observation in different ranges.

The subjects of statistical mechanics and quantum theory appear to forbid the notion of an intrinsic exact law; probabilities occur even in the most precise statements imaginable. They were originally derived from attempts to relate the phenomena of heat and radiation to those of classical mechanics and electromagnetism. Boyle showed that for most gases the volume of a given specimen at constant temperature is inversely proportional to the pressure. Charles showed that at given pressure gases have almost equal coefficients of thermal expansion, as if the volumes would vanish if the gases were cooled to $-273°$ C. The two laws together imply the relation

$$pv = R'\vartheta,$$

where p is the pressure, v the volume per unit mass, ϑ the temperature measured from $-273°$ C, and R' a constant for a given gas. For typical gases (say with densities of the order of 10^{-3} gm./cm.³) the law is correct to about 1 part in 1000.

The idea that matter consists of discrete particles, much smaller than the smallest that we can distinguish directly, goes back to classical times (Democritus, Lucretius), but the first scientific application was by Gassendi, followed by Hooke and D. Bernoulli. There were curious delays in drawing conclusions that now seem almost obvious, and serious attention to the atomic hypothesis had to wait for Dalton's laws of chemical combination.

The eighteenth-century writers proceeded on the following lines. Suppose that a gas contains ν particles of mass m per unit volume, and consider a small piece of its containing surface, with area A, which we shall treat as plane. Consider a thin layer in contact with A. A particle with velocity w towards the boundary will strike the boundary in a short time dt if its original distance was less than wdt. If the number in a range dw and in a volume element $d\tau$ is $\nu f(w)\, dw\, d\tau$, where $\int_{-\infty}^{\infty} f(w)\, dw = 1$, the number striking the boundary in time dt is

$$A\nu w\, dt \,.\, f(w)\, dw. \tag{1}$$

Each molecule of velocity w has its velocity normal to the boundary reversed on striking the boundary and therefore communicates momentum $2mw$ to the boundary. The momentum communicated to the boundary per unit area per unit time is therefore

$$p = 2m\nu \int_{0}^{\infty} w^2 f(w)\, dw = m\nu c_w^2, \tag{2}$$

say.

But the unit volume contains mass νm, so that

$$v = 1/\nu m, \tag{3}$$

and hence
$$pv = c_w^2. \tag{4}$$

The pressure is the same in whatever position the boundary is placed, including places within the containing vessel, which suggests that the expectation of w^2 is the same for components in all directions. (It does not prove it because the presence of an instrument for measuring the pressure might alter the distribution of velocity.) But the uniformity of the measured p and ϑ through the chamber suggests that c_w^2 is the same for all positions and all directions, and is identical with $R'\vartheta$. We can then write c for c_w. Thus for a given gas the temperature is proportional to the mean square velocity.

The velocities indicated are rather high. For air in ordinary conditions p is about 10^6 dynes/cm.^{3} and the volume of 1 gm. is about 10^3 cubic centimetres. Thus c is of order 3×10^4 cm./sec., comparable with the velocity of sound. In ordinary experiments the velocities of the containers are much less than c and can be regarded as small disturbances of the general agitation.

The theory so far merely provides a possible model; c has to be calculated from p and v, but all the argument says is that *if* we suppose that a gas consists of many particles in a state of agitation, then the mean square velocity in each component must be given by (4). Little further progress was possible until Dalton stated the laws of chemical combination, which indicated that each chemical element entered into combination only in exact multiples of a definite mass. His form, however, left an ambiguity. Thus he wrote the formula for a molecule of water as HO; but it was possible on his data that the smallest quantity of hydrogen as such might be any submultiple of that denoted by H in his formula, and similarly for oxygen. Avogadro noticed that if the R' in the Boyle-Charles law was multiplied by the Dalton mass the products were integral multiples of a constant, and this fact led to his famous rule. If the true molecular mass is m, we can write

$$R' = k/m, \tag{5}$$

where k is the same for all gases at low densities, and then the Boyle-Charles law will read

$$mv \cdot p = k\vartheta, \tag{6}$$

where k is Boltzmann's constant. But mv is the volume per molecule, and the rule can be read: at given temperature and pressure, equal volumes of all gases contain the same number of molecules. The number in a gram molecule is Avogadro's constant, N: the gas constant R is Nk.

Avogadro's rule can be regarded as a more precise way of stating Dalton's rules, so far as these concern gases; the extra precision being obtained by reducing the arbitrariness of the possible values of the mass to be associated with a molecule. It led to the results that the molecular formulae of hydrogen, oxygen, and water must be written H_2, O_2, H_2O instead of H, O, and HO. So far it need be regarded as only a definition of a molecule. Its great practical advantage was that ratios of molecular weights could now be definitely assigned by comparison of densities in the gaseous state. What remained doubtful was whether this definition of a molecule agreed with the older idea of a molecule as the smallest quantity of a substance that could exist. If we do identify the Avogadro molecule with the freely moving molecules of (4) we can write (6) as

$$k\vartheta = mc^2, \tag{7}$$

so that the temperature is identified with a constant multiple of the kinetic energy per molecule.

There was a surprising delay before this last step was taken. Heat was generally regarded as a substance capable of being transferred from one body to another, but incapable of being created or destroyed. Even Carnot in his derivation of the so-called second law of thermodynamics from the theory of heat engines adopted this view; he never drew the conclusion that mechanical energy could be converted to heat or vice versa, in spite of the existence of D. Bernoulli's interpretation of the Boyle-Charles law. Rumford had actually made the substance theory of heat very unplausible, but when Joule made actual measures of the generation of heat by doing work against friction his results met wide opposition. Kelvin and Clausius were ultimately convinced, and restated Carnot's principle in a form that explicitly regarded heat as a form of energy.

There is one serious difficulty in the application of (7) rather than (2). (2) contains only measured quantities. (7) contains also the unknown mass of a molecule. If we accept the constancy of k, studies of the properties of substances in the gaseous state will give the ratios of the molecular masses, not their actual values. The practical procedure in chemistry is to take the mass of the oxygen atom as 16, thus defining a new mass unit whose ratio to the gram was still to be found.

The first attempts to determine m came through the phenomena of viscosity. So far we have been considering gases in stationary enclosures. But when a gas or liquid is pushed through a tube the pressures at the ends differ; the velocity distribution is the same over all cross-sections, so that there is no acceleration. Thus the force due to the difference of pressure is balanced by a tangential force over the walls of the tube. This arises as follows. Suppose that the main flow is in the direction of x, but that the mean velocity is a function of y. The mean square of the velocities in the y direction is c^2. Let the molecules be spheres of diameter σ; then if the centres of two approach to a distance σ there is a collision and the momenta are shared between the two. In a distance dy, per unit surface in the xz plane, there will be $\nu \, dy$ molecules. Then a molecule departing from $y = 0$ toward positive y will have a chance of about $\pi \nu \sigma^2 \, dy$ of hitting another in distance dy. Thus the expectation of the distance it will travel before colliding is of order $1/\nu \sigma^2$. Now the x momentum is mu and on an average the excess over the mean at the place where the collision takes place is $\dfrac{m}{\nu \sigma^2} \dfrac{du}{dy}$. The number of molecules crossing unit area per unit time is of order νc; thus the rate of transfer of momentum is of order $\dfrac{mc}{\sigma^2} \dfrac{du}{dy}$. If this is the explanation of viscosity in a gas, the coefficient of viscosity is

$$\eta \eqsim mc/\sigma^2. \tag{8}$$

Here m and σ are independent of pressure and temperature; c is proportional to $\vartheta^{\frac{1}{2}}$. Thus η should be proportional to $\vartheta^{\frac{1}{2}}$ and independent of pressure. Independence of pressure is well verified, the variation with temperature not so well, since η generally varies faster than $\vartheta^{\frac{1}{2}}$ but not so fast as ϑ. The latter point can be explained if the molecules are elastic.

Now if the substance in the solid state has density ρ, we can suppose that m and σ retain the same values, and the density will be of order m/σ^3. Thus comparison of the viscosity in the gaseous state with the density in the solid state will give estimates of m and σ. They are very small by ordinary standards; the mass of the hydrogen molecule is about 3×10^{-24} gm., and the molecular diameters of the simpler elements and compounds are of order 10^{-8} cm.

The theory was first developed to some extent by J. J. Waterston, whose paper was refused publication until it was rediscovered by Lord Rayleigh. Maxwell gave a very full development and many extensions followed. The more detailed theory of viscosity, with the related phenomena of heat conduction and diffusion, is complicated and depends on the shapes of the molecules and the actual law of force between them. The most accurate determination of Boltzmann's constant k depends on Millikan's determination of the charge on an electron. In the electrolysis of water the mass of hydrogen liberated per unit charge transmitted can be found, and hence the charge on the electron determines the mass of an atom of hydrogen.

In the rough theory outlined above actual values have several times been replaced by expectations, and this procedure is necessary in any treatment. It would be formally possible for the answers to be completely wrong. For instance, in the account of viscosity it would be possible for the molecules to be arranged at intervals of order σ in rows parallel to the y-axis; the number of particles per unit volume could still be ν if the distance between the rows was of order $1/\nu\sigma^2$. Then a molecule would travel only a distance of order σ before a collision instead of $1/\nu\sigma^2$. It has really been assumed that positions in planes more than σ apart are independent. Again, it is assumed that at a collision the molecule considered shares its x momentum, with the molecule it hits, and that the two then proceed independently to collisions with others. Thus the properties inferred for a gas are essentially statistical; they depend on the use of expectations on various hypotheses of randomness instead of the true values. In a region expected to contain about 10^6 molecules it would be expected that actual numbers will fluctuate by about 10^3, and the mean square velocity by 10^{-3} of its mean value. Since an actual chamber will contain 10^{20} or more molecules, the fluctuations will be of order 10^{-4} of the mean values if we consider samples whose volumes are of order 10^{-12} of that of the whole chamber. Thus if the hypothesis of randomness is correct, expectations will be indistinguishable in practice from observed values.

Thus the kinetic theory of gases deals essentially with fluctuating motions that we do not know in detail. It uses the principles of classical mechanics, but uses them to derive relations between statistical properties, and hence is essentially an application of probability theory. †

10·1. It must be remarked that current presentations pay very inadequate attention to probability theory, and in consequence their explanations are unsatisfactory. An essential point is that when

---

† An excellent article by C. G. Darwin (*Proc. Roy. Soc.* A, **236**, 1956, 285–96) deals with the topics of this section more fully. The fullest discussion is by S. Chapman and T. G. Cowling, *The Mathematical Theory of Non-Uniform Gases*, Cambridge University Press (3rd ed. 1970).

a chamber is filled with a gas, that gas has at first a systematic motion away from the orifice, far greater than would be likely to occur by ordinary fluctuation. Collisions deflect the particles and gradually eliminate the systematic motion. In part it is converted into heat and the changes are part of the subject matter of thermodynamics. The primary problem is to see how this can come about.

There are two standard approaches. The first is by way of the H theorem of Boltzmann. This considers the velocity components of two molecules before and after collision. The positions and velocities before collision are treated as independent, and hence the joint probability (for six co-ordinates and six velocity components) is found by multiplication. It is then shown that the changes of velocity at collision are such that the expectation of the change of a certain quantity H is negative. Assuming that it does not tend to $-\infty$, it will be expected to tend to a steady value, and this state is found to be such that any molecule is equally likely to be anywhere in the chamber and the probability of each component of its velocity must satisfy the normal law of error about zero; in this context it is called Maxwell's law. In answer to this argument it must be pointed out that the problem is to explain why a distribution, *not* initially satisfying Maxwell's law, should approach that law. If there are systematic motions initially, the fact that one molecule has a positive x component of velocity is some evidence that it lies in a stream toward positive x, and therefore that a molecule near it will also lie in the stream and have a positive x velocity. Hence the hypothesis that the velocities of colliding molecules have independent probability distributions is false, and the argument breaks down at the start.

The other argument is by way of the ensemble of Willard Gibbs. Instead of considering one system it considers an infinite number of systems in similar enclosures, with the same number of particles, but otherwise arranged at random. Average properties and average (mean square) fluctuations over systems with equal total energy are then considered, and the Maxwellian distribution is derived as an approximation. Here there are two objections. First, the notion of averages in an infinite set of systems introduces ratios of infinite numbers; but such ratios are totally indeterminate. The arithmetic of finite numbers cannot be extended to infinite numbers, and any attempt to do so leads to inconsistencies. (See Appendix I.)

Secondly, the hypothesis of randomness subject to the given values of the number of particles and the energy begs the whole question. For a gas newly admitted to an enclosure has motions that are not random in this sense, and the argument does not even attempt to explain why these should die out.

10·2. A possible treatment is as follows. We use rectangular co-ordinates q and momenta p for every particle; if there are N particles, the state at any instant is specified by $3N$ co-ordinates and $3N$ momenta. The information about the initial state can be expressed by saying that the probability distribution of the q's and p's at time o is given by

$$f(q_1 \cdots q_{3N}, p_1 \cdots p_{3N}) \, dp_1 \cdots dp_{3N} \, dq_1 \cdots dq_{3N}, \tag{1}$$

which we can write briefly as

$$f(q, p) \, dq \, dp. \tag{2}$$

Now for precise values of q and p the values q', p' at time t are determinate, and by a theorem of Liouville, for any conservative dynamical system the Jacobian

$$\frac{\partial(q', p')}{\partial(q, p)} = 1. \tag{3}$$

The probability that q', p' lie in specified small ranges at time t is simply the probability that q, p lie in the corresponding ranges at $t = 0$; but this is $f(q, p) \, dq \, dp$, which, by (3), is the same as $f(q, p) \, dq' \, dp'$.

If we consider the $6N$ dimensional element expressed by $dq \, dp$, and regard it as moving with the time, (3) implies that it always maintains the same volume. But since p has values differing by dp within the element, $\dot{q}$ has values differing by quantities of the order of dp/m. Similarly since $\dot{p} = -\partial H/\partial q$, which in general varies with q, $\dot{p}$ has values within the element differing by $O(dq)$. Thus, given enough time, the differences of the extreme values of the co-ordinates and momenta within the element may become large. Thus, though the element always keeps the same volume, it changes drastically in shape, becoming longer in some directions and shorter in others.

In actual conditions of observation an instrument is of finite size and is set up at a fixed place. Now for a fixed rectangular element

$dq'\,dp'$, with t large, the corresponding values at $t=0$ do not form a rectangular element or anything like one; any single rectangular element at $t=0$ of the same dimensions gives an elongated element at t, whose width is much less than that of $dq\,dp$ in some directions at least. Hence the probability that q', p' will be within a specified element at time t comes by adding contributions from many different ranges of q, p, corresponding to different values of f. If it is true that for any set q, p the $6N$ dimensional path will ultimately pass arbitrarily near any given q', p', this implies that the probability density will tend to constancy; that is, if t is large enough, and we take rectangular elements of given size, the probability that q', p' lie in a specified element tends to a constant multiple of $dq'\,dp'$. In fact there are other conditions, such as the constancy of energy, that impose restrictions on the admissible values of q' and p', given q and p, but these can be allowed for without difficulty. The rest of the theory can then be developed as usual.†

Many accounts claim the result as a consequence of Liouville's theorem. This is nonsense, as was pointed out long ago by Jeans and Fowler. The constancy of the element of volume in (q, p) space is only the first step in the argument; Fowler called the further conclusion a 'pious hope'.‡ The distortion of the element with time is essential to the argument.

The essential point is that events on a sufficiently small scale cannot be observed in detail, and their tendency is to convert any systematic motion into random motion, otherwise heat. This is a case of the second law of thermodynamics.

Theorems that, for a given system started off in any way, any other state satisfying the energy condition will ultimately be approached arbitrarily closely, have been proved for many types of system and are known as ergodic theorems. The late Sir R. H. Fowler remarked to me that my paper was the first to bring out clearly what these theorems had to do with statistical mechanics.§ For a gas the condition is plausible because the initial information is not precise enough to predict even one collision, and successive collisions can transmit momentum throughout the volume.

† H. Jeffreys, *Proc. Roy. Soc.* A, **160**, 1937, 337–48.
‡ *Statistical Mechanics* (1936), p. 14.
§ The fundamental probability theorem is in *Phil. Mag.* (7) **33**, 1942, 815–31, which gives earlier references. See also H. and B. S. Jeffreys, *Methods of Mathematical Physics* (1950), 163.

For a solid the problem is more complicated because each atom has a standard position, about which it can oscillate. Equalization of temperature between a gas and its container is possible because collisions of the gas molecules with the boundary set up oscillations in the solid. The motion of the solid can be analysed, to a first approximation, into normal modes, and if transfer of energy is possible between any two modes, the probability tends to such a form that the mean kinetic and the mean potential energy in any mode are equal to mc^2 for the gas. The probability distribution of either follows the rule

$$P(dE \mid H) = e^{-x} dx, \quad \text{where} \quad x = E/k\vartheta. \tag{4}$$

It is not, however, obvious that transfer between any pair of modes is possible. It is plausible, since the analysis into normal modes depends on neglecting square terms in the equations of motion. If these terms are applied as a correction they introduce reactions between the normal modes. Verification of the rule (4) is evidence for the existence of suitable means of transfer of energy; this cannot usually be proved theoretically. In fact (4) is an approximation. The theory leads to the result that for given total energy E, with n degrees of freedom, the probability that m of them carry energy in a range dE_1 is proportional to

$$E_1^{\frac{1}{2}m-1}(E - E_1)^{\frac{1}{2}(n-m)-1} dE_1.$$

The expectation of E_1 is exactly mE/n. For n large and m moderate the probability of E_1 is closely concentrated about this value. Thus energy is distributed in approximate proportion to the number of degrees of freedom. But the energies of different degrees of freedom are not independent, being restricted by the condition that their sum is E. (4), if interpreted as exact, implies that one degree of freedom could absorb an arbitrarily large amount of energy, and therefore more than the total energy.

10·21. A system may have properties that can be measured in small samples of its members. For some such properties the probability for the whole system can be derived from those of the samples and no other information is relevant. Such measures are called sufficient statistics, and by a theorem of Pitman and Koopman† the probability of a measure x must be of the form

$$f(x, \alpha_1...\alpha_m) = \phi(\alpha_1...\alpha_m)\, \psi(x)\, \exp \Sigma u_s(\alpha)\, v_s(x). \tag{1}$$

ϕ is fixed by the condition that the sum of f over all possible values of x must be 1. The factor in the likelihood that depends on both the observations and the parameters is $\Sigma S u_s(\alpha)\, v_s(x)$ (Σ being a sum over α, S one over x). Thus $S v_s(x)$ contains all the information in the data relevant to $u_s(\alpha)$. Further, since the posterior probability is nearly proportional to the likelihood, we shall have

$$P(\alpha_s \mid \theta H) = \exp S u_s(\alpha) v_s(x) / \Sigma \exp u_s S v_s \tag{2}$$

for any parameter that tends to a constant value over the system; the summation in the denominator is over all possible values of α.

The mean square velocity of agitation over a system can be interpreted as proportional to the temperature, and this can be measured over different parts of the system.

Now comparing this with 10·1 (4), if we put $v(x) = dE$, $\alpha = -1/k\vartheta$, we have agreement. The argument in the first place concerns an induction from many samples, possibly each containing 10^{20} molecules, to an experiment possibly containing 10^{24}. The comparison however shows that there is no contradiction in extending the law down to 1 molecule.

Application of the principle of equipartition of energy to radiation in fact led to an absurd consequence, which was resolved only by the quantum theory. This difficulty, noticed by Rayleigh, concerned an enclosure containing no matter but with a heated boundary, so that the enclosure is filled with radiation. The radiation consists of electromagnetic waves, and in classical theory they can be arbitrarily short. If free transfer of energy can take place, equipartition will be set up. Let the speed ‡ of an oscillation be σ. Then there are in general many possible modes of vibration with speeds in a range $d\sigma$; for large σ the number is in fact proportional to $\sigma^2 d\sigma$. If the average energy was the same for every mode, the total energy would contain a factor $\int_0^\infty \sigma^2 d\sigma$, which is infinite. Thus there is no possible

---

† Jeffreys, *Theory of Probability* (1967), p.168.
‡ In tidal theory, if an oscillation is expressed by sin γt, γ is called the speed. This term is shorter than others that are in use.

steady state for a given finite total energy, and this is in contradiction with experiment.

It appears that the contradiction can arise only through the hypothesis that general interchange of energy can take place. In other words, the larger values of σ (i.e. the shorter wave-lengths) are excluded in some way from the rule of equipartition. It seems queer that Rayleigh's argument was accepted as a deduction from the laws of classical physics, which it certainly is not. A general ergodic argument would be needed to complete it. The electromagnetic equations are exactly linear, and provide no means of interchange between modes. Reactions with the container offer more hope, but the argument should be put the other way round: equipartition does not hold for radiation at high frequencies, and therefore there is some restriction on transfer of energy.

There is much other evidence tending in the same direction. The ratio of the specific heats of a gas at constant pressure and constant volume, $\gamma = c_p/c_v$, is $5/3$ for gases whose molecules consist of a single atom, $7/5$ for diatomic atoms. If equipartition holds, it can be shown that

$$\frac{c_p}{c_v} = 1 + \frac{2}{3+n}, \tag{3}$$

where n is the number of degrees of freedom corresponding to rotations and internal vibrations. For any molecule we should expect three degrees of freedom to correspond to rotations, and for an elastic molecule there might be an indefinitely large number of vibrations. Thus γ might be near 1, and even for a gas of inelastic monatomic molecules we should have $n = 3$, $\gamma = 4/3$. There is disagreement with observation for monatomic gases unless $n = 0$; that is, the rotations and internal vibrations are not excited at all. This might be possible for rigid, smooth spherical atoms, though it would be straining the hypotheses to the extreme. But for diatomic molecules, besides the possible rotations and internal vibrations of the two atoms, three co-ordinates are needed to express the position of one atom relative to the other. Even if we restrict ourselves to these they would contribute at least 3 to n, and since the radial component must for stability be associated with a radial force, which would contribute to the energy, the natural number would be 4, making γ equal to $9/7$ instead of $7/5$. In fact to get $7/5$ we must have

$n = 2$. This can be taken to mean that the two angular displacements occur but that the radial one does not. Thus the kinetic theory of gases needs severe qualification even with regard to some of its most striking successes. It was noticed, however, that if equipartition took place, the rotations of the atoms in diatomic molecules would have to be far faster than those of the orbital revolutions, and the speeds of the internal vibrations would have to be faster still. The radial vibration, if it existed, might have a higher speed than the rotations. Thus the excluded motions are all possible instances of particularly short periods. There is some confirmation in the fact that at very low temperatures γ for diatomic gases rises towards its value for monatomic ones, as if for low energies the molecular rotations also are not fully excited.

Similar anomalies were found for solids. Equipartition leads to the conclusion that the specific heat per molecule (molecular rotations again neglected) should be the same for all solid elements, and this is true for most elements at ordinary temperatures. But at low temperatures the specific heat falls off greatly, as if it was tending to 0 at the absolute zero of temperature.

These relations are summed up in what is called the Nernst heat theorem or the third law of thermodynamics. Landé discusses this in detail.†

In any case it looks as if the more rapid motions are excluded from the rule of equipartition, and as if the restriction becomes more serious at low temperatures, when the general motion tending to excite them becomes less violent. In spite of the criticism of Rayleigh's argument made above, the conclusion that equipartition is not general is verified.

Most of this information was available to Planck, who created the quantum theory. The clearest experimental evidence was for radiation in an enclosure, the so-called black-body radiation. For this the actual distribution of energy with respect to wave-frequency was measured. The result is that the intensity between speeds σ and $\sigma + d\sigma$ is

$$E_\sigma \, d\sigma = \frac{A\sigma^3}{e^{\hbar\sigma/k\vartheta} - 1} \, d\sigma, \tag{4}$$

where A is a constant; $\hbar$ is a new universal constant, equal to Planck's constant h divided by 2π. k is Boltzmann's constant,

$$k = R/N. \tag{5}$$

† A. Landé, *New Foundations of Quantum Mechanics* (1965), Cambridge University Press.

So long as $\hbar\sigma$ is small compared with $k\vartheta$, (4) is nearly proportional to $\sigma^2\vartheta\,d\sigma$; thus the rule of equipartition gives the right form at low speeds (long waves). But if $\hbar\sigma/k\vartheta$ is large it approximates to $A\sigma^3 e^{-\hbar\sigma/k\vartheta}\,d\sigma$, and the actual energies in the vibrations of high speeds are far less than equipartition would imply. Planck showed that the form (4) is consistent with the hypothesis that for any speed σ the only possible energies are o, $\hbar\sigma$, $2\hbar\sigma$, ... For $\hbar\sigma/k\vartheta$ large the great majority of modes of speed σ are not excited at all. This result is derived only from the behaviour of radiation, but it is qualitatively just the sort of thing needed to account for the departures of gases and crystals from other properties predicted by the rule of equipartition. On the other hand transfer of energy between modes of vibration remains an essential part of the theory, and if only a discrete set of values are possible it follows that any transfer must be discontinuous. This is contrary to any form of classical mechanics, which makes all transfers continuous. (Impulses appear in classical mechanics, but only as approximations to very rapid continuous processes.)

10·22. *Entropy.* Suppose that a system of N members has n possible states, of which the probabilities are p_i; the number of members in the ith state is N_i, which is not necessarily the expectation Np_i. The probability of the set is

$$P(N_i) = \frac{N!}{N_1!\ldots N_n!}\, p_1^{N_i}\ldots p_n^{N_n}. \tag{1}$$

The logarithm is close to

$$\tfrac{1}{2}\log(2\pi)^{n-1} + \{(N+\tfrac{1}{2})\log N - N\}$$
$$- \Sigma\{(N_i+\tfrac{1}{2})\log N_i - N_i\} + \Sigma N_i \log p_i. \tag{2}$$

We first want its maximum subject to $\Sigma N_i = N$. The large terms in (2) are, with $m_i = Np_i$,

$$-\Sigma(N_i \log N_i - N_i) + \Sigma N_i \log (m_i/N) \tag{3}$$

and $\Sigma N_i = N$ is invariable. Introducing an undetermined multiplier λ and varying the N_i we have the conditions for a stationary value

$$\{-\log N_i + \log (m_i/N) + \lambda\}\, \delta N_i = \text{o}, \tag{4}$$

for all i, whence $$N_i = \frac{m_i}{N} e^\lambda,$$

and by summation $$N = e^\lambda, \quad N_i = m_i. \tag{5}$$

The departure of (3) from its maximum is then

$$-\Sigma N_i \log (N_i/m_i). \tag{6}$$

If we put $$N_i = m_i + \alpha_i m_i^{\frac{1}{2}} \tag{7}$$

and expand the logarithm, this reduces to

$$-\tfrac{1}{2}\Sigma\alpha_i^2 + O(m_i^{-\frac{1}{2}}) = -\tfrac{1}{2}\Sigma \frac{(N_i - m_i)^2}{m_i}. \tag{8}$$

This is Pearson's $-\tfrac{1}{2}\chi^2$, which is important in significance tests.

It has been pointed out, notably by Fowler in his *Statistical Mechanics*, that Stirling's formula used in (2) assumes that the N_i are large, and is sometimes used for $N_i = 0$ or 1. As a matter of fact it is not badly wrong for $N_i = 1$. (Fowler has methods for approximating at a later stage, which avoid the difficulty.) But the expansion of the logarithm can make much larger errors, failing completely if $\alpha_i m_i^{-\frac{1}{2}} \geqslant 1$. The difference (4) is free from this difficulty. With the sign changed it has the property that any approach to the most probable state increases it, and it therefore has the essential property of entropy in thermodynamics. Various formulae for this exist, some of them severely criticized by Landé.

Variations of $\log P(N_i)$ can be written as those of

$$-\Sigma N_i \log N_i/p_i$$

which retains the property that it increases with departure from expectation. In either form, if two systems are allowed to mix, the increase of entropy is to be found by comparing both with the mixture when the latter has reached a steady state.

This form has some similarity to the invariant

$$J = \Sigma(p_i' - p_i)\log(p_i'/p_i)$$

for the comparison of two probability distributions, which I have found useful in stating prior probabilities.

An approach similar to this is given by B. Mandelbrot.† He

† *Ann. Math. Stat.* **33**, 1962, 1021–38.

traces, incidentally, the Pitman–Koopman theorem back to Poincaré (1896) and Szilard (1925, 1929). Another interesting account, relating thermodynamics to quantum theory, is by R. T. Cox.†

10·3. Extensions of the theory to matter constitute quantum mechanics. Light emitted by atoms was found to have wave-lengths characteristic of the atoms to high accuracy. The natural interpretation, when electrons were discovered, was that these moved in periodic orbits, and that the motion of these charged particles produced an electromagnetic field with the orbital periodicity. However Larmor pointed out that this radiative field would carry away energy and the electron should gradually collapse on the nucleus. This would imply a continual increase in the frequency of the radiation, which does not happen. Bohr explained this by supposing that the energy of an atomic orbit and also the angular momentum were capable only of discrete values, and the differences of the energy in different orbits corresponded, according to the rule $E = \hbar\sigma$, to the frequencies of the light that the atom was capable of emitting. Larmor's difficulty disappears if states arbitrarily close to the actual one are impossible and if radiation occurs only when there is some disturbance. The theory was found to be capable of accounting for the peculiar behaviour of specific heats, and achieved numerous other successes. Perhaps the clearest evidence was the photoelectric effect. Light of speed σ falling on a metal produced no effect but heating when σ was less than a characteristic value σ_0; for $\sigma > \sigma_0$ electrons were ejected with energies $\hbar(\sigma - \sigma_0)$. It appeared that the light came in packets of energy $\hbar\sigma$, each of which could be absorbed by only one atom, which itself took up energy $\hbar\sigma_0$ and the rest was radiated. Which atom it would be was evidently a matter of chance.

However, it was incomplete. Apparently any transition between two states should be possible and give a characteristic radiation; but the observed radiations included only some of those apparently possible, though some additional ones could be produced by special conditions, notably magnetic fields. Further, the theory did not predict the intensities of the spectral lines that do occur. This was remedied by the newer quantum mechanics, different forms of

† *Statistical Mechanics of Irreversible Change* (1955), Johns Hopkins Press.

which were stated nearly simultaneously by Heisenberg, Dirac, and Schrödinger. It has been shown that all give equivalent answers for any observationally verifiable result, though the terminologies are very different. I shall speak mostly in terms of Schrödinger's system, which is the most widely used.

In classical mechanics, if we use Cartesian co-ordinates, there are a kinetic energy

$$T = \Sigma m_\alpha \dot{x}_i^2 \tag{1}$$

and a potential energy

$$V = -W = V(x_1 \ldots x_{3N}). \tag{2}$$

We introduce the momenta

$$p_i = \frac{\partial T}{\partial x_i} \tag{3}$$

and the Hamiltonian function

$$H = T + V. \tag{4}$$

Then we use (3) to eliminate the $\dot{x}_i$ from H, thus expressing H as a function of the p's and q's. It can then be proved that in any actual motion of the system

$$\frac{dx_i}{dt} = \frac{\partial H}{\partial p_i}; \quad \frac{dp_i}{dt} = -\frac{\partial H}{\partial x_i}. \tag{5}$$

These are Hamilton's equations. Further, if we consider the Hamilton-Jacobi differential equation

$$\frac{\partial S}{\partial t} = -H\left(\frac{\partial S}{\partial x_i}, x_i\right), \tag{6}$$

this has a solution containing $3N + 1$ adjustable constants, say

$$S = S(x_i, \alpha_1 \ldots \alpha_{3N}, t) + \alpha_{3N+1}. \tag{7}$$

If we put

$$p_s = \partial S / \partial x_s, \quad \beta_r = \partial S / \partial \alpha_r \quad (r \leqslant 3N) \tag{8}$$

and treat α_r, β_r as constants, it can be shown that x_s and p_s so defined satisfy Hamilton's equations and in fact represent the most general motion of the system.

The solutions contain $6N$ adjustable constants, which can be taken to be the co-ordinates and momenta at time o.

In the actual solution of the problem of the motion of a planet $\alpha_1, \alpha_2, \alpha_3$ arise as the energy, the resultant angular momentum, and the component of angular momentum in a given plane; $\beta_1, \beta_2, \beta_3$ are the longitude at a standard time, the longitude of perihelion measured from the node on a fixed plane, and the longitude of the node.

In a more general system we can define the Poisson bracket of any two functions a, b of the co-ordinates and momenta as

$$\{a, b\} = \sum_i \left(\frac{\partial a}{\partial x_i} \frac{\partial b}{\partial p_i} - \frac{\partial a}{\partial p_i} \frac{\partial b}{\partial x_i} \right)$$

and it has the property that if q_r', p_s' are chosen, functions of q_r, p_s, so that

$$\{q_r', q_s'\} = 0, \quad \{p_r', p_s'\} = 0$$
$$\{q_r', p_r'\} = 1, \quad \{q_r', p_s'\} = 0 \quad (r \neq s)$$

then q_r', p_r' satisfy the Hamiltonian equations of motion. Such a set of transformed variables is called *canonical*. There are standard methods in dynamical astronomy for discovering variables of this type for use in successive approximation. q_r, p_r with this property are called *conjugates*. The Cartesian co-ordinates and momenta are conjugates.

An important property is that the rate of change of any quantity is its Poisson bracket with the Hamiltonian:

$$\frac{da}{dr} = \{a, H\}.$$

Poisson brackets play an important part in Dirac's theory.

In the solution for N particles α_{6N+1} disappears from Hamilton's equations, leaving $6N$ relevant ones.

In quantum theory we are concerned with possibly discontinuous motions, which also have a further random element besides that considered in classical statistical mechanics. (5) will not necessarily be true; but any change must be such as will leave (5) true as an approximation for phenomena on a large scale. We no longer necessarily speak of a particle as having an exactly calculable position; instead its position has a certain probability distribution, which changes with the time. This may be regarded as a wave.†

---

† If a man sings a note, lasting, say, 0·2 sec., the air motion at a listener is oscillatory for about 0·2 sec., performing something of the order of 50 vibrations in this interval. The air movement as a whole travels out with uniiorm velocity, like a particle, but a gramophone record records the details of the oscillation. Diffraction of light pulses and of electrons shows such wave properties. A difference is that whereas the note may be almost equally audible to hearers in any direction, the light pulse or the electron can be absorbed by only one atom, and any appearance of uniformity arises only through random occurrences for a large number. It is as if each note was heard by only one listener.

(To identify the particle with the wave is fallacious.) Now since S in an actual orbit increases uniformly with the time like $-Et$, where E is the total energy, we may look for a function

$$\psi \doteq e^{\lambda S}, \qquad (9)$$

where λ is purely imaginary and ψ satisfies a linear differential equation. H can still be written down as a function of the xs and ps. We take only one particle for simplicity. Then

$$H = \frac{1}{2} \frac{p^2}{m} + V, \qquad (10)$$

$$\frac{\partial S}{\partial t} = -\frac{1}{2m}\left(\frac{\partial S}{\partial x}\right)^2 - V, \qquad (11)$$

and

$$\frac{\partial S}{\partial t} \doteq \frac{1}{\lambda \psi} \frac{\partial \psi}{\partial t}, \qquad (12)$$

$$\frac{\partial^2 S}{\partial x^2} \doteq \frac{1}{\lambda \psi} \frac{\partial^2 \psi}{\partial x^2} - \lambda \left(\frac{\partial S}{\partial x}\right)^2, \qquad (13)$$

$$\frac{1}{\lambda \psi} \frac{\partial \psi}{\partial t} \doteq -\frac{1}{2m}\left(\frac{1}{\lambda^2 \psi} \frac{\partial^2 \psi}{\partial x^2} - \frac{1}{\lambda} \frac{\partial^2 S}{\partial x^2}\right) - V. \qquad (14)$$

Now S is of dimensions ML^2/T, and so is Planck's constant. For (9) to have meaning, then, λ must be of dimensions $1/h$ and can be treated as large. Since each differentiation of ψ brings in a factor λ, the terms in $\partial^2\psi/\partial x^2$ are not small, but the term in S is. If we drop this term we have a linear equation for ψ; and in particular if we take

$$\lambda = i/\hbar, \qquad (15)$$

we have

$$\frac{\hbar}{i} \frac{\partial \psi}{\partial t} = \frac{\hbar^2}{2m} \frac{\partial^2 \psi}{\partial x^2} - V\psi. \qquad (16)$$

This is Schrödinger's equation. Its extension to any number of particles moving in three dimensions is immediate.

If in (16) we substitute (9), we are led back approximately to the Hamilton-Jacobi equation in problems where S/h is large. If then S is treated as in (8), the whole of ordinary dynamics comes out as an approximation.

All we have done so far, however, is to find a less convenient way of expressing classical dynamics. The special properties of (16) appear when we look for solutions in specially concentrated fields.

If V represents an attractive force toward a point, say $V = \frac{1}{2}kx^2$, and we look for a solution of the form

$$\psi = e^{i\sigma t} f(x),$$

we find that in general all solutions of the differential equation are unbounded; but for special values of σ there is one bounded solution. These values are integral multiples of a constant.[†] For an electron in the neighbourhood of a positively charged nucleus there are bounded solutions for a discrete set of energies, provided the energy is not too great. If it is larger any value of the energy is possible—in classical terms, if the electron is moving fast enough,' so that the orbit is hyperbolic. The distinction between bound and free electrons thus arises naturally. This is just what is needed to explain the sharpness of atomic spectra. The essential point is the prediction that an atom can exist in a set of discrete states. Consequently any transition between these states must be discontinuous, and the essential feature of quantum phenomena emerges naturally.

The introduction of ψ invites the question: what is ψ? Here the leaders in the subject are in disagreement. To Heisenberg and Dirac, Schrödinger's equation seems to be only a way of discovering the possible states of a system. Schrödinger himself regards ψ as an ultimate reality; where we might speak of N particles in a universe, Schrödinger would say that the universe consists of ψ waves in $3N$ dimensions. Born's interpretation, however, is most widely used. If the specification of the position of the system needs N co-ordinates ψ is a function of just these variables and the time. Then in N-space $\psi\psi^* d\tau$ is the probability that the position of the system at time t lies in the element $d\tau$. Here ψ^* is the conjugate complex to ψ, so that the probability is always real and non-negative.

For different values of σ $(= -E/\hbar)$, say $\sigma_1, \sigma_2 \ldots \sigma_m$, there will be solutions $\psi_1 \ldots \psi_m$. These can be shown to be orthogonal in the sense that for different σ_r, σ_s

$$\int \psi_r \psi_s^* \, d\tau = 0 \qquad (17)$$

† The deduction from his equation for various models was made by Schrödinger, *Ann. Phys. Lpz.* **79**, 1926, 361–76 and 489–527. See also H. and B. S. Jeffreys, *Methods of Mathematical Physics*, Sections 23.07 and 23.08. In these references use is made of the asymptotic approximations to the wave functions. An apparently simpler but less satisfactory method appears in many textbooks.

integration being through all co-ordinate space, and if different ψ_r correspond to the same σ it is still possible to form linear combinations of them with the same property. Further, any solution of (16) can be expressed as a sum

$$\psi = \Sigma a_s \psi_s. \tag{18}$$

If a continuous set of σ is involved the sum will be replaced by an integral. Then

$$\psi \psi^* = \Sigma \Sigma a_r \psi_r a_s^* \psi_s^*. \tag{19}$$

However this is dependent on t. If, as we are assuming, V is a function of the co-ordinates and not of the time, the function S must be of the form $-Et+$ a function of x, α. But if ψ is a linear combination of ψ_m, ψ_n, with different time factors, log ψ cannot be even approximately of this form. Equation (16) is linear but (11) is not. The suitable interpretation is that different ψ_m, ψ_n correspond to different states of the system, and the propositions that they are correct are exclusive alternatives, each giving its separate contribution to the probability density. If we average (19) over a time long compared with the periods we obtain an expression of the form $\Sigma c_m \psi_m \psi_m^*$, where the c_m are positive quantities adding up to 1; c_m expresses the probability that the system is in the mth state. If both c_m and c_n are non-zero it must not be interpreted as saying that the system is partly in the mth and partly in the nth state; it means that it may be in either of those states but we are not certain which.

On this interpretation a particle is not a wave. It is ψ that has wave properties, and it is only an intermediary in calculating the probability that the particle will be in a particular element of space.

10·31. However, this statement is incomplete, because a probability is always relative to a set of data, and the data are not specified. They must in any case be somewhat peculiar. C. G. Darwin† has given a solution for the motion of an electron in a uniform field, and his

† *Proc. Roy. Soc.* A, **117**, 1927, 258-93.

answer is equivalent to the statement that the probability distribution for its position at time t is

$$P(dx \mid t, H) = \frac{1}{\sqrt{(2\pi)} \, (\sigma^2 + \hbar^2 t^2/4m^2\sigma^2)^{\frac{1}{2}}} \exp - \frac{(x - a - ut - \frac{1}{2}gt^2)^2}{2(\sigma^2 + \hbar^2 t^2/4m^2\sigma^2)} \, dx,$$

$$(1)$$

where a, u, σ are constants. This is exactly the distribution that would hold if x at $t = 0$ is given to be $a \pm \sigma$ and the momentum $p = m\dot{x}$ at $t = 0$ is $mu \pm \hbar/2\sigma$, these uncertainties being independent, and if the particle moves with strictly constant acceleration g. Then H can be taken to represent the information

$$P(dx_0 dp_0 \mid t = 0, H)$$

$$= \frac{dx_0}{\sqrt{(2\pi)} \cdot \sigma} \exp\left(-\frac{(x_0 - a)^2}{2\sigma^2}\right) \frac{dp_0}{\sqrt{(2\pi)} \, \hbar/2\sigma} \exp\left(-\frac{(p_0 - mu)^2}{\hbar^2/2\sigma^2}\right), \quad (2)$$

and for $t \neq 0$, $\qquad x = x_0 + p_0 t/m + \frac{1}{2}gt^2, \quad p = p_0 + mgt.$ $\qquad\qquad (3)$

The product of the standard errors of x_0 and p_0 at $t = 0$ is always $\frac{1}{2}\hbar$. If we transform to variables x and p instead of x_0 and p_0, (2) gives a joint probability distribution for q and p at any time t, namely

$$P(dx \, dp \mid t, H)$$

$$= \frac{1}{\pi\hbar} \exp\left[-\frac{1}{2}\left(\frac{x - a - pt/m + \frac{1}{2}gt^2}{\sigma}\right)^2 - \frac{1}{2}\left(\frac{p - mu - mgt}{\hbar/2\sigma}\right)^2\right] dx \, dp.$$

$$(4)$$

Integration with regard to p gives (1) again, as it should. The presence of a term in xpt in the exponent shows that when $t > 0$ the uncertainties are correlated; the discriminant of the exponent is

$$\begin{vmatrix} \dfrac{1}{\sigma^2} & -\dfrac{t}{m\sigma^2} \\[3mm] -\dfrac{t}{m\sigma^2} & \dfrac{t^2}{m^2\sigma^2} + \dfrac{1}{\hbar^2/4\sigma^2} \end{vmatrix} = \frac{4}{\hbar^2}.$$

The area of the element in x, p-space therefore does not change with the time; but it does change in shape. This is just the feature that appeared in the dynamical theory of gases and gave the explanation of equipartition.

10·32. The quantum theory has become entangled with epistemological theories, most of which are extremely doubtful, and in particular with an unfortunate form of phenomenalism. Darwin's solution for an electron in a uniform field serves to illustrate some of them. On the face of it, it gives the probability distribution of x at any time, and with a natural extension gives also the joint probability distribution of x and p at any time. What are the data? Apparently the values of x and p at $t = 0$, with standard errors σ and $\hbar/2\sigma$, subject to the condition that there is no other observation between 0 and t, since extra information would alter the probability distribution. The solution thus depends on the uncertainties of the initial values being related in a particular way; the more accurately a co-ordinate is known at $t = 0$, the less accurately is the corresponding momentum known. This is one case of Heisenberg's uncertainty principle. This has been misinterpreted. The idea that physical measurements were not made with absolute accuracy produced a sensation in 1928, and many philosophers, and even theoretical physicists, still appear to think that it was an astonishing new discovery. Some remained (and remain) prepared to deny it at all costs. It was, of course, a platitude to anybody that had handled actual observations. What Heisenberg pointed out was that any attempt to measure the position of an electron would require the use of radiation of very short wave-length. This, if it was affected by the electron at all, would produce a reaction (the Compton effect), and if the position of the electron was determined with uncertainty σ the momentum would be changed by something of order h/σ. Even the fact that making an observation may alter the quantity that we are trying to measure, or related quantities, was already familiar; a voltmeter alters the electric current, an anemometer affects the local wind and hence the distribution of pressure in its neighbourhood. The real point is that physicists had been accustomed to suppose that, given detailed knowledge of the system and the actual values of the variables at one instant, the exact values at any other time could be calculated, and that, though there were practical restrictions on the accuracy of knowledge of the variables, there need be no restriction in principle. Heisenberg's argument appeared to show that there was an intrinsic uncertainty due to the effect of the observation itself.

However, it seems to me that this was not adequately proved, and that further conclusions have been drawn that are certainly false. In classical mechanics the uncertainty of the data had to be allowed for if the predictions were to be valid, even though the laws themselves were supposed to be exact. We spoke of actual values existing even though we did not claim to know them exactly, and the actual uncertainties were far greater than those contemplated by Heisenberg. This situation is hardly altered by his argument; it would still be possible to maintain that the co-ordinate and momentum had exact values at any instant, but that they changed discontinuously at an observation.

The real point is different. We might regard the X-ray as a particle; if it passes the electron on the side of negative x we might expect it to give an impulse toward positive x, and conversely, and hence that the errors in x and p at $t = 0$ would be closely correlated. But according to 10·31 (2) they are not correlated at all. It is this independence of the errors that lies at the root of the matter.

Later writers have gone further still and maintained that it is meaningless even to speak of co-ordinates and momenta as having simultaneous probability distributions. As we have seen, however, Darwin's solution from wave mechanics is certainly consistent with a simultaneous probability distribution. Further, the whole of classical mechanics depends on the existence of such distributions. If we knew only the position of a body at an instant, and nothing at all about its momentum, we could predict nothing at all about its position at any other instant. Quantum mechanics, if it is to be comprehensive, must be in a position to derive the classical equations of motion as approximations valid for systems containing many atoms; and, however it is to be done, some variables corresponding to the co-ordinates and momenta must persist. To deny that they can have a simultaneous probability distribution is to say that quantum mechanics can never explain why classical mechanics gives the right answers for the motion of the planets.

The immediate point is that the Darwin electron is consistent with the supposition that the motion is deterministic in the classical sense for $t > 0$, and that the uncertainties of x and p are connected in a special way, at least one of them being less than could be attained by any actual observation. The data, therefore, must be a particular and much idealized type of observation at $t = 0$.

The natural extension of the result would be that motion

ordinarily follows classical rules, but that there are discontinuous changes at intervals; and these have the property that they give changes of a pair of conjugate variables with independent probability distributions such that the product of the expectations of their squares is $\frac{1}{4}\hbar^2$. Independence of different pairs of conjugate canonical variables would be expected, and this would apparently have some relation to the assumption in quantum theory that the corresponding matrices or operators commute. See also 10·35.

Two general probability theorems relating to series of events suggest further directions of inquiry. The first is due to M. Fréchet.

10·33. Consider the phase space in the sense of 10·2 (1) (2) divided into elements of equal volume and suppose that the initial data make the probability that the system is in the ith set of elements equal to x_i at time o. Suppose also that the probability that it will be in the ith set at time t, given that it is in the jth set at time o, is a_{ij}. Then by the product and addition rules the probability, on the initial data, that it will be in the ith set at time t is

$$y_i = \sum_j a_{ij} x_j. \tag{1}$$

We have in any case $\sum_i x_i = 1$, $\sum_i a_{ij} = 1$; and we can also suppose the elements so chosen that $\sum_j a_{ij} = 1$. The probabilities that the system will be in the ith set at times $2t$, $3t$... are found by repetitions of the same transformation. This is what is called a Markov chain; the theory is mostly due to M. Fréchet.† Consider the set of equations

$$\sum_j a_{ij} \theta_j = \lambda \theta_i. \tag{2}$$

In general these are consistent if λ has one of a certain set of values, called characteristic values or eigenvalues, for each of which the θ_i are in a particular set of ratios; and a general set of x_i can be expressed as a linear combination of these sets, say

$$x_i = \sum_r \alpha_r \theta_{ir} \tag{3}$$

† In Borel's *Traité du Calcul des Probabilités*, **1**, Fasc. 3, 1938; a fuller account than the present one is in H. and B. S. Jeffreys, *Methods of Mathematical Physics*, 4·16. See also the last footnote on p. 220.

and y_i will be $\sum_r \lambda_r \alpha_r \theta_{ir}$. The result of p applications will be $\sum_r \lambda_r^p \alpha_r \theta_{ir}$. It is shown in the works cited that one value of λ is always 1; the others are in general complex but all have moduli $\leqslant 1$. If all a_{ij} with $i=j$ are different from zero it is impossible for any λ other than 1 to have modulus 1; that is, if whatever state the system is in there is a non-zero probability that it will stay in that state. In that case the probabilities of all the states will tend to constants. On the other hand if there is a λ with $|\lambda|=1$ but $\lambda \neq 1$ we can choose the θ of largest modulus, say $|\theta_1|=R$, and then some other θ must have modulus R. If no two θs are equal this can be satisfied only if for some i all $a_{ij}=0$ except for one value of j, which makes $a_{ij}=1$; then for this j the $a_{jk}=0$ for all k except one, which makes $a_{jk}=1$, and so on. If the set closes in m steps all the λ are of the form $\exp(2\pi i r/m)$. Thus any λ not 1 but having modulus 1 specifies a succession of determined steps; no branching is possible. See also an example at the end of p. 240.

There can be more than one such cycle. Then the matrix a_{ij} is in diagonal block form and each block gives an independent set of solutions.

The above argument supposes a finite set of possible elements of phase space, all of non-zero volume. If it can be extended to continuous distributions, as seems reasonable, either all $|\lambda_r| \neq 1$ have $|\lambda| < 1$, for which the y_i, or some functions of them, will tend to constant values, or at least one $|\lambda_r|=1$ and $\lambda_r \neq 1$, and some function of them describes a definite orbit in the classical sense. Such a part may be an atom or even a molecule. If there are interactions the corresponding $|\lambda|$ may become <1, but if these are small or rare they may be only a little under 1, and periodic orbits can be used as a first approximation. In a crystal, where interactions are large, we can retain the form by taking components of displacement and momentum in the normal co-ordinates as q and p instead of the rectangular components for separate particles. It seems interesting in any case that we can get as far as this from probability theory alone.

This argument has supposed the a_{ij} constants, satisfying $\sum_i a_{ij}=1$, and that the elements are all of equal volume in phase space. In this case for $\lambda=1$ the θ_i can all be equal; for if in (2) we put $\lambda=1$

we have
$$\sum_j a_{ij}\theta_j = \theta_i$$

and if all the θ_i are equal this reduces to $\sum_j a_{ij} = 1$, which is satisfied for every i. Unless there are other solutions for $\lambda = 1$, therefore, the system will tend to a set of periodic solutions added to one that makes all the elements of phase space equally probable; and even small interactions will make the periodic solutions disappear. We have equipartition again. This clearly cannot be extended to quantum theory, where if the elements are taken small enough most of them would be impossible. Even in classical dynamics, if there are interactions, the field of force will depend on which of the elements are occupied by particles, and this will depend on the time. Thus it is doubtful whether it is legitimate to take the a_{ij} as constants.

A related difficulty seems to have been met by Eddington and by Hartree. Eddington speaks of a 'rigid field theory'. An observed particle disturbs its surroundings and the disturbance reacts on the particle. He states that the ordinary method in physics is to treat the surroundings as rigid and to transfer the consequences of this error to properties of the particle. Most of his results are applications of this principle. (He may or may not have applied it correctly.) In his introduction he appears to be appealing to statistical mechanics. There have been at least four attempts to explain his principles, but I cannot understand any of them any better than the original.

Hartree† speaks of a 'self-consistent field', which is essentially an approximation independent of the time. As a first approximation the charge density in an atom is taken to be proportional to $\psi\psi^*$, calculated for a positive charge on the nucleus, according to an early view of Schrödinger. This is averaged with regard to the angular co-ordinates. The potential is revised to take it into account and the equation is solved afresh. The calculation is repeated with the revised ψ, until it converges. The agreement with observed energy levels is surprisingly good—for helium well under 1 per cent. Such a process would presumably give constant values for the a_{ij} if these were evaluated.

† D. R. Hartree, *Proc. Camb. Phil. Soc.* **24**, 1928, 89–132, and later papers in *Proc. Roy. Soc.* A. *Reports on Progress in Physics*, **11**, 1947, 113–43; *The Calculation of Atomic Structures*, Wiley, 1957.

In an atom with more than one electron we should expect that, as for the gravitational interaction between planets, there will be perturbations of each electron by the others. But these will in general have periods incommensurable with the orbital periods, and it is reasonable that for long-term behaviour most of the effects will cancel.

Fréchet's analysis and that in our book consider only finite sets of alternatives, but the extension to continuous sets involves using integrals. If in phase space in the sense of 10·2 (1) and (2) the probability distribution at time 0 is

$$P(dq\,dp\,|\,h) = F(q,p)\,dq\,dp \qquad (4)$$

and the values at time t are q', p', we can write

$$P(dq'\,dp'\,|\,h) = G(q',p')\,dq'\,dp', \qquad (5)$$

$$P(dq'\,dp'\,|\,qph) = K(q',p',q,p,t)\,dq'\,dp' \qquad (6)$$

and by the addition and product rules

$$G(q',p') = \int\int K(q',p',q,p,t)\,F(q,p)\,dq\,dp. \qquad (7)$$

If there are N particles in the system, we must remember that q, p are shorthands for $q_1 \ldots q_{3N}$ and $p_1 \ldots p_{3N}$. The whole of the dynamics of the system is contained in K. Repetition to a further time $2t$ gives a similar relation, and the probabilities at successive dates rt are related in a Markov chain.

We must have

$$\int\int F(q,p)\,dq\,dp = 1\,;\quad \int\int G(q',p')\,dq'\,dp' = 1, \qquad (8)$$

$$\int\int K(q',p',q,p,t)\,dq'\,dp' = 1 \qquad (9)$$

and in accordance with our method in 10·2 we can suppose the sizes of the elements $dq\,dp$ so chosen that also

$$\int\int K(q',p',q,p,t)\,dq\,dp = 1. \qquad (10)$$

Now if there is a set of functions $\theta_l(q,p)$ such that any physically possible $F(q,p)$ can be linearly expressed in terms of them, and such that

$$\iint K(q',p',q,p,t)\theta_l(q,p)\,dq\,dp = \lambda_l\theta_l(q',p') \qquad (11)$$

we can take

$$F(q,p) = \sum_l A_l\theta_l(q,p), \quad G(q',p') = \sum_l A_l\lambda_l\theta_l(q',p') \qquad (12)$$

and at rt a factor λ_l^r occurs.

To determine the constants A_l in (12) we should require the θ_l to form a complex orthogonal set,

$$\iint \theta_l\theta_m^*(q',p')\,dq'\,dp' = 0 \quad \text{if } l \neq m. \qquad (13)$$

In my 1942 paper quoted on p. 220 I treated the continuous case by reducing it to the finite one for which I related the condition $|\lambda_l| = 1$ for all l to reversibility of path and relativity of time.

If θ_l is constant, (11) reduces to $1 = \lambda_l$. Hence 1 is always a solution. There may be many such solutions.

Consider a solution with $|\lambda| = 1$, $\lambda \neq 1$. Let $|\theta|$ take its greatest value R for a pair of values q_1, p_1. Then the right side of (11) has modulus R. On the left $|\theta| \leqslant R$, and since θ is in general complex and K integrates to 1, the left side is a weighted mean, and will have modulus $< R$ unless $\theta = 0$ except for one pair, or possibly a set, of values of q and p, for which it is equal to R. $|\theta|$ may have a maximum with regard to p and be constant with regard to q, and in general there will be exceptional cases if the maximum is taken for more than one set of values.

On p. 237 we supposed that no two θs are equal. An interesting example when two θs are equal has been given me by Mr M. J. Pelling. If the matrix is

$$\begin{pmatrix} 0 & 0 & \frac{1}{2} & \frac{1}{2} \\ 0 & 0 & \frac{1}{2} & \frac{1}{2} \\ \frac{1}{2} & \frac{1}{2} & 0 & 0 \\ \frac{1}{2} & \frac{1}{2} & 0 & 0 \end{pmatrix}$$

the characteristic values are 0, 0, 1, -1, and the corresponding solutions for θ are

λ	0	0	1	-1
θ_1	1	0	1	1
θ_2	-1	0	1	1
θ_3	0	1	1	-1
θ_4	0	-1	1	-1

Thus $|\theta|$ for $\lambda = -1$ is 1 for all i. If we took x_1 and $x_2 \neq 0$, x_3 and $x_4 = 0$, we should get $y_1 = 0$, $y_2 = 0$, $y_3 = y_4 = \frac{1}{2}(x_1 + x_2)$; and further values depend only on $x_1 + x_2$ and not on either separately.

The condition for $|\theta|$ to take its largest value could be satisfied if it was constant for some function of q, p and a maximum for another; for instance an angular momentum and an orbital angle.

In the case of a finite set of alternatives there is at least one solution with $\lambda = 1$; but if there is a positive probability for every state that the system will stay in that state up to time t, then every other solution has $|\lambda| < 1$.

The immediate conclusion is that probability theory does not exclude exact orbits.

10·34. *The Cramér–Yaglom theorem*.† It is often convenient in dealing with periodic motions in time or space to combine $\cos \gamma t$, $\sin \gamma t$ and $\cos \kappa x$, $\sin \kappa x$ into $\exp(i\gamma t)$, $\exp(i\kappa x)$, but if so we shall have to consider complex values of the variables. Now if there are n variables, functions of t, say $\xi_1(t)$, $\xi_2(t) \dots \xi_n(t)$, we may consider the cross covariance between times t and s

$$B_{jk}(t,s) = E\xi_j(t)\xi_k^*(s) \tag{1}$$

* denoting the complex conjugate and E the expectation. The point of using the complex conjugate is that

$$E(A e^{i\gamma t} A' e^{i\gamma s}) = AA' e^{i\gamma(t+s)} \tag{2}$$

whereas $\qquad E(A e^{i\gamma t} A'^* e^{-i\gamma s}) = AA'^* e^{i\gamma(t-s)}. \tag{3}$

The second depends only on the time interval $s - t$; the first contains the absolute time.

† H. Cramér, *Ann. Math.* (2) **41**, 1940, 215–30; Cramér and M. R. Leadbetter, *Stationary Random Functions and Stochastic Processes*, Chapter 7. Wiley, 1967. A. M. Yaglom, *Stationary Random Functions*, trans. R. A. Silverman. Prentice-Hall 1962.

Now if B_{jk} depend only on the interval $t - s = \tau$ we may say that the set ξ is in a stationary state, and write the covariances as a matrix

$$\mathbf{B}(t, s) = \mathbf{B}(t - s) = \mathbf{B}(\tau). \tag{4}$$

Then

$$B_{jk}(\tau) = E\xi_j(s + \tau)\xi_k^*(s), \tag{5}$$

$$B_{jk}(-\tau) = E\xi_j(s - \tau)\xi_k^*(s) = E\xi_j(s)\xi_k^*(s + \tau)$$

$$= B_{kj}^*(\tau) = B_{jk}^\dagger(\tau), \tag{6}$$

where the $\dagger$ denotes the transpose of the complex conjugate. But the elements of the covariances at intervals $-\tau$ and τ are the same in a long-continued process. Thus the covariances $\mathbf{B}$ constitute a Hermitian matrix.

Also we can write

$$\mathbf{B}(\tau) = \int e^{i\omega\tau} \, d\mathbf{F}(\omega), \tag{7}$$

where F may be called the spectral distribution of the variables. Then

$$\mathbf{B}^*(\tau) = \int e^{-i\omega\tau} \, d\mathbf{F}^*, \tag{8}$$

$$\mathbf{B}^*(-\tau) = \int e^{i\omega\tau} \, d\mathbf{F}^*, \tag{9}$$

$$\mathbf{B}^\dagger(-\tau) = \int e^{i\omega\tau} \, d\mathbf{F}^\dagger, \tag{10}$$

But by (6),

$$\mathbf{B}^\dagger(-\tau) = \mathbf{B}(\tau). \tag{11}$$

whence F is also a Hermitian matrix not involving τ.

Now a Hermitian matrix can always be reduced to diagonal form by a unitary matrix $\mathbf{l}$ satisfying

$$\mathbf{l}\mathbf{l}^\dagger = \mathbf{1} \tag{12}$$

that is, there is an $\mathbf{l}$ such that

$$\mathbf{l}^\dagger \mathbf{B}\mathbf{l} = \boldsymbol{\mu},$$

where $\boldsymbol{\mu}$ is a diagonal matrix with all its elements real.

The characteristic values of a unitary matrix have modulus 1. Hence they may be identical with λ_r, which in turn may be of the form $\exp iE_r t/\hbar$, where E is an energy. Also, since energy and time are conjugate variables in Hamiltonian theory, there is a suggestion that this can be extended to all co-ordinates and momenta, giving another factor $\exp(\pm ipq/\hbar)$.

For real variables the expectations form a symmetric matrix, and there is 'detailed balancing'; that is, the expectations of transfers from state A to state B and from B to A are equal in the steady state. However the main point is that the importance of Hermitian and unitary matrices is essentially a consequence of the restriction to steady states, but with the extension that periodic orbits can be included.

As the components of μ are real they cannot of course be identical with the complex λ considered above. The **B,** like the a_{ij}, are supposed independent of the time and therefore involve a self-consistent field approximation.

Landé is a sensible realist; he rejects the conversion of a particle to a wave and back, as in most treatments of quantum theory. He also makes much more extended criticisms of the metaphysics of the theory than I have given above. He emphasizes in particular Duane's explanation of the reflexion of electrons by a crystal. Most quantists use the above mysterious conversion, and in fact are proud of it. Duane's explanation is that a periodic structure with wave-length l (with harmonics l/r) can undergo only changes of momentum rh/l. When an electron meets it, by the conservation of momentum, it also will undergo changes of momentum rh/l; and this accounts for the facts. Incidentally this has some similarity with Eddington's outlook.

There may appear to be an inconsistency between Fréchet's and Yaglom's theorems, and some doubt about their applicability. The arguments concern systems with a finite set of alternatives, but are applied to continuous sytems. There is an extensive theory of solutions of homogeneous integral and differential equations, and it is known† that the characteristic solutions can form a finite set, an infinite discrete set, a continuous set, or a continuous set together with a discrete one, depending largely on whether a coefficient becomes infinite anywhere. Most of the results for a finite set do hold for an infinite one except that sums may need to be replaced by integrals.

The apparent inconsistency between the two theorems is in the treatment of periodic solutions. According to Fréchet's theorem no such solution can exist unless there is a set of values $a_{ij}, j = i$,

† E. C. Titchmarsh, *Eigenfunction Expansions*, chapters 5, 16. Oxford University Press 1961.

all zero. This means that for some initial states the system is certain to move out of the state in the first interval, and in Yaglom's model if $\tau \neq 0$, B_{ij} $(i=j)$ should be 0. But if this was so every state would be initially unoccupied. This anomaly arises however through a reversal of the order of limiting processes. If the interval in Fréchet's theorem is α, the correlation between values in the same interval would be 1. The use of non-zero intervals has introduced a spurious discontinuity, which could be avoided by a suitable way of making the sizes of the time and space intervals tend to zero suitably.

When the αs and βs are used as variables in Fréchet's theorem most of the a_{ij} will be zero or small. The exceptions will be for those capable of simultaneous transitions. The possible amounts of the transitions will be fixed by the quantum conditions. There is no inconsistency between the theorems and the idea of exact orbits between transitions at random intervals of time, which would smooth the probabilities. It has been usual to regard the transitions as instantaneous, but Fairlie's work (see p. 248) yields an actual expression for the rate of transition. Such expressions, based on ψ alone, had already been given for many problems but only for small disturbances.

By the Cramér-Yaglom theorem B is a sum (or integral) of terms of the form $F(\omega)\exp(i\omega\tau)$, and the exponential factor is naturally interpreted as the λ in Fréchet's theorem.

10·341. In classical mechanics any motion (at any rate in a bounded region) could be expressed in an infinite Fourier series. When a gas is freshly admitted to a chamber and then left alone, the representation can be in periodic terms. The original systematic motion away from the orifice is represented by a large number of these being in the same phase. As time goes on the phases become randomized on account of incommensurability of periods.

Landé uses a principle similar to one used by R. P. Feynman.†
If A, B, C refer to states at moments of time we have in general

$$P(ABC\,|\,h) = P(A\,|\,h)P(B\,|\,Ah)P(C\,|\,ABh) \qquad (1)$$

and
$$P(AC\,|\,h) = \sum_B P(A\,|\,h)P(B\,|\,Ah)P(C\,|\,ABh) \qquad (2)$$

† *Rev. Mod. Phys.* **20**, 1948, 367–87.

and also $$= P(A \mid h) P(C \mid Ah). \tag{3}$$

Hence $$P(C \mid Ah) = \sum_B P(B \mid Ah) P(C \mid ABh). \tag{4}$$

They replace this by

$$P(AC) = \sum_B P(AB) P(BC). \tag{5}$$

$P(C \mid ABh)$ has essentially been replaced by $P(C \mid Bh)$. This is valid only if A is irrelevant to C given B. If the Ps were functions of position only, knowledge of where the system is at B could tell us very little about that at C, but in addition the position at A provides information about the velocities between A and B and therefore between B and C. If the Ps also refer to the momenta the result might be true. However there is a further complication, since the quantities related in their usage are supposed to be observed values, and an observation at B in general disturbs the system. $P(AC \mid h)$ does not mention B and assumes no disturbance between A and C. But $P(AB)$ and $P(BC)$ will be different according as the measurements are made just before or just after the disturbance at B, and in either case there is no apparent reason why (5) should be true.

Landé uses a principle similar to one used by R. P. Feynman.† states $A_1, A_2 \dots$ or $B_1, B_2 \dots$. He appeals to two experimental facts: (1) if the A test has been applied and the particles in A_1 are separated, further applications of A yield entirely A_1; (2) if the B test is applied to particles that have already undergone an A test, they are again split up, but the probability of a particle in state A_i going into state B_j is equal to that of one in B_j going into A_i. This appears to be of fundamental importance. What I am doubtful about is whether the above argument applies in general. What I think is intended is that the quantities measured at A, B, C are only some of those correct immediately after A, B, C; these undergo changes at the measurements, specified by suitable a_{ij}, but between the measurements all the canonical parameters remain constant. In fact we are back at exact orbits but have information now about what actually happens at transitions.

The quantities measured are usually those corresponding to the αs of classical theory (for instance energy and angular momentum). Then the assumption that the measurement at A gives no information relevant to the probabilities at C, given B, refers only to these parameters, and the probabilities concerned are integrals in phase space with regard to the others.

A point often made is that in a quantum transition the probabilities of the Bs are independent of the history before A. But if the states expressed by ψ really involve independent uncertainties of the initial states, as appeared in our discussion of the Darwin electron, these would be expected to combine in successive transitions, and each scatter would increase. If however there are exact orbits this would not occur. It has appeared that the expectations yielded by ψ do in fact satisfy the classical equations of motion exactly, and there is really no objection to supposing that these are exact solutions. The probabilities of transitions remain a further problem. This takes us back to before 1925, and the meaning of ψ remains a mystery.

The triumph of wave mechanics was that, whereas earlier theories predicted frequencies, it also predicted amplitudes. Curiously, though Landé's account of Duane's theory explains deflexions, amplitudes are not mentioned.

Landé has two further assumptions: (3) if after passing B the A test is applied again, the original distribution is recovered; that is, the matrix for the transition $B \to A$ is the reciprocal of that from A to B; (4) if tests A, B, C, D are applied in turn, the result is the same as if they are applied in the order A, C, B, D; that is, the matrices of transformation at B and C commute. Now it is not true that all symmetrical matrices commute, and this gives a further restriction. But combining (1) with (3) we appear to have that the reciprocal of a transformation matrix is the transpose; that means that these matrices are also orthogonal. But this cannot be true for orthogonal matrices with no negative elements unless they are diagonal with all diagonal elements 1.

10·342. Corresponding to $\psi(q)$ there is a function of momentum $\phi(p)$ such that

$$\phi(p) = (2\pi\hbar)^{-\frac{1}{2}} \int e^{-ipq/\hbar} \psi(q) \, dq \qquad (1a)$$

and

$$\psi(q) = (2\pi\hbar)^{-\frac{1}{2}} \int e^{ipq/\hbar} \phi(p) \, dp. \qquad (1b)$$

The probability density for p is taken to be $\phi\phi^*$. Orthogonal solutions for ψ lead to orthogonal ones for ϕ.

In Fréchet's theorem (p. 236), if applied to quantum theory, the x_i, y_i, would be probabilities in elements $dq\,dp$ in phase space, and we may hope that the orthogonal eigenfunctions θ_{ir} can be expressed in terms of ψ and ϕ. We should look for functions $F(q, p)$ constructed from different solutions of a given Schrödinger equation such that the integral of $F_{ab}(q, p)$ with regard to p gives $\psi_a\psi_b^*$ and the integral with regard to q gives $\phi_a\phi_b^*$. This suggests taking

$$F_{ab} = (2\pi\hbar)^{-\frac{1}{2}}\psi_a\phi_b^* \exp\left(-ipq/\hbar\right). \tag{2}$$

But for a general F to be expressible compactly in terms of such functions we should require complex orthogonality in the sense that $\iint F_{ab}F_{a'b'}^*\,dq\,dp = 0$ unless $a = a'$, $b = b'$. This is satisfied in most cases, but if λ is a solution of 10·33 (2), λ^* is another, and the corresponding F will be F^*. Then orthogonality for such pairs of solutions would require $\iint (F_{ab})^2\,dq\,dp = 0$, which appears to be false in general. I got as far as this after my 1942 paper but could make no further progress.

A considerable advance is made by Moyal and Bartlett,[†] following results of Kermack, McCrea and Wigner. If we write

$$\phi(p) = (2\pi\hbar)^{-\frac{1}{2}}\int \psi(q)\,e^{-ipq/\hbar}\,dq \tag{3}$$

the joint probability density of q and p is taken to be

$$F(q,p) = \frac{1}{2\pi}\int \psi^*(q - \tfrac{1}{2}\hbar\xi)\,e^{-i\xi p}\,\psi(q + \tfrac{1}{2}\hbar\xi)\,d\xi$$

$$= (2\pi\hbar)^{-\frac{1}{2}}\exp\left(\frac{1}{2}\frac{\hbar}{i}\frac{\partial^2}{\partial q\partial p}\right)\left[\psi^*(q)\,\phi(p)\,e^{ipq/\hbar}\right]. \tag{4}$$

It is shown that if F is real and positive at $t = 0$ it is so for all t. Further, if ψ and ϕ are taken to be derived from two solutions of Schrödinger's equation, say ψ_k, ϕ_l, the resulting F_{kl} have the properties of the **B** obtained above. They have the further properties that if two of them differ in at least one suffix and the complex conjugate of one is taken, the integral of the product over all space is zero; they form a Hermitian matrix; and they form a complete

† Proc. Camb. Phil. Soc. **45**, 1949, 99–124, 545–53.

set, that is, any function of the ps and qs (if reasonably well behaved) can be expressed as a linear function of them. The expectations of the qs and ps satisfy the classical equations of motion, so that the connexion between quantum and classical mechanics is made.

Further developments are given by D. B. Fairlie.[†] In particular Schrödinger's equation is found explicitly for a steady state and a possible modification for time dependence, taking account of an external disturbance, not necessarily small, is given. The theory is given as for one degree of freedom but could obviously be extended to any number of rectangular co-ordinates.

There is one apparent discrepancy between this work and the theory of 10·33. It appeared that in an undisturbed system there would be definite orbits, one specified by each F. Moyal and Bartlett appear to find that this is true only when the potential is a quadratic function of the co-ordinates. This would still leave some indeterminacy in, for instance, the motion of an electron about a nucleus.

I have tried to locate the maximum moduli of their eigenfunctions, but they appear to be very complicated. We may note however that if the λ and λ^* solutions of Fréchet's theorem are replaced by their real and imaginary parts, these are orthogonal. Professor Littlewood gave me the following proof.

We use the allied Fourier integral. In the Fourier integral

$$\frac{1}{2\pi} \int_{-\infty}^{\infty} f(x) e^{i\kappa x} dx \qquad (5)$$

the results of separating the real and imaginary parts may be denoted by $\phi_1(\kappa)$, $\phi_2(\kappa)$. Then if

$$\int_{-\infty}^{\infty} |f(x)|^2 dx$$

exists consider, with $\epsilon > 0$,

$$\int_{-\infty}^{\infty} e^{-\epsilon u^2} \phi_1(\kappa) \phi_2(\kappa) d\kappa$$

$$= \iint f(x) f(y) dx dy \int_{-\infty}^{\infty} e^{-\epsilon u^2} \cos \kappa x \sin \kappa y \, d\kappa.$$

[†] *Proc. Camb. Phil. Soc.* **60**, 1964, 581–6.

$\cos \kappa x \sin \kappa y$ is the difference of two sines and the integral with regard to κ is zero. Make $\epsilon \to 0$ and use the principle of bounded convergence.† Then it follows that

$$\int \phi_1(\kappa)\, \phi_2(\kappa)\, d\kappa = 0.$$

The allied Fourier integral (and the analogous series) have many physical applications, and there seems to be some possibility that this may be another.

The analysis of 10·33 concerns the probabilities at later times given those at t_0. It does not consider the possibility that there may have been a disturbance actually at t_0. Now we saw under the Darwin electron that the solution was consistent with uniform acceleration given the position and momentum at t_0; that the expectations of q and p given by a particular ψ solution do in fact satisfy the classical equations of motion; and that the choice of a particular F_r with $\lambda \neq 1$, $|\lambda| = 1$ implies an exact relation between q and p at any time with those at t_0. Since the expectations are in fact unique it appears that all these hints can be combined into one statement: that between random disturbances the system moves according to the classical equations; ψ is only indirectly connected with the actual motion, but in some way the possible ψ solutions identify what orbits are possible. The crucial point is that at a transition the disturbances of a pair of conjugates are independent.

It is sometimes argued that quantum theory does not deal with individual motions at all, but is concerned only with states that are steady in the same sense as those of a gas in an enclosure at constant temperature. However, this is unsatisfactory; apart from the fact that it again blocks any attempt to explain classical mechanics in terms of quantum mechanics, it neglects all quantum phenomena that do not concern steady states. When a gas is heated it begins to radiate at once; it does not wait till it has reached a steady temperature and then begin to radiate.

10·35. *The uncertainty principle.* Using the probability density $\psi\psi^*$ as in 10·3 in one-dimensional coordinate space, we have that if

$$\int \psi\psi^*\, dx = 1$$

† H. and B. S. Jeffreys, *Methods of Mathematical Physics*, Note 1.116a. The proof is due to A. S. Besicovitch.

the expectation of x is

$$\bar{x} = \int x\psi\psi^* \, dx \tag{1}$$

and that of the conjugate momentum is

$$\bar{p} = -i\hbar \int \psi^* \frac{\partial \psi}{\partial x} \, dx \tag{2}$$

and these satisfy the classical equations of motion. Further, as in 10·342, the probability density in momentum space is $\phi\phi^*$.

The following is a proof of the uncertainty principle for the case when $\bar{x}$ and $\bar{p}$ are zero. The general result, that the product of the expectations of $(x-\bar{x})^2$ and $(p-\bar{p})^2$ is not less than $\frac{1}{4}\hbar^2$ can be proved similarly. We use a form of Fourier's theorem, and assume that $\psi(x)$ is such that $\int|\psi|^2 dx$, $\int|x^2|\psi|^2 dx$, $\int|\psi'|^2 dx$ are all finite (the termini all being $\pm\infty$); with $p = \hbar\omega$, if

$$\chi(\omega) = \hbar^{\frac{1}{2}}\phi(p) = \frac{1}{\sqrt{(2\pi)}} \int e^{-i\omega x}\psi(x) \, dx, \text{ then } \psi(x)$$

then
$$\psi(x) = \frac{1}{\sqrt{(2\pi)}} \int e^{i\omega x}\chi(\omega) \, d\omega. \tag{3}$$

We have, if $\psi = \mu + i\nu$

$$\int|\psi|^2 dx = x|\psi|^2 - \int 2x(\mu\mu' + \nu\nu') \, dx \tag{4}$$

the integrated part vanishing; now if λ is real

$$\int [\lambda x(\mu - i\nu) + (\mu' - i\nu')][\lambda x(\mu + i\nu) + (\mu' + \nu')] \, dx \tag{5}$$

is positive or zero, and hence

$$\left[\int x(\mu\mu' + \nu\nu') \, dx\right]^2 \leqslant \int x^2(\mu^2 + \nu^2) \, dx \int (\mu'^2 + \nu'^2) \, dx \tag{6}$$

that is,

$$\left[\int|\psi|^2 dx\right]^2 \leqslant 4\int x^2|\psi|^2 dx \int |\psi'|^2 dx. \tag{7}$$

But

$$\psi'(x) = \frac{1}{\sqrt{(2\pi)}} \int i\omega e^{i\omega x}\chi(\omega) \, d\omega \tag{8}$$

which is the transform of $i\omega\chi(\omega)$. Hence by Parseval's theorem

$$\int |\psi'|^2 dx = \int \omega^2 |\chi(\omega)|^2 d\omega \tag{9}$$

and finally

$$\left[\int |\psi|^2 dx \right]^2 \leqslant 4 \int x^2 |\psi(x)|^2 dx \int \omega^2 |\chi(\omega)|^2 d\omega. \tag{10}$$

Putting $\omega = p/\hbar$ and remembering that $\int |\psi|^2 dx = 1$, we have that the product of the expectations of x^2 and p^2 is always $\geqslant \frac{1}{4}\hbar^2$. The inequality becomes an equality only if ψ' is proportional to $x\psi$; then ψ must be of the form $A \exp(-\frac{1}{4}x^2/\sigma^2)$, and the Darwin electron at $t = 0$ is the only solution in the case when $\bar{x} = 0$ and $\bar{p} = 0$.

10·4. I do not think that any of the standard arguments against determinism in quantum theory are conclusive against determinism holding in the classical sense, when the meaning of this is understood. However, there are quantum phenomena that make it very difficult to maintain. The most striking is the α-particle counts given by Rutherford and Geiger (p. 48). These represent a random distribution in the sense that, subject to the parameter in the Poisson law, the chance of an emission in any interval is independent of occurrences in other intervals, as if all atoms of the specimen were independent. Again, consider a series of radioactive elements, in which three consecutive members are A, B, C, where B has a much longer life than A or C. Collect the B from A over a time-interval τ short compared with the average life of B. The most striking event in the history of any of these B atoms is the disruption of the A atom that produced it. If determinism holds, we should expect none of the B atoms to break up and produce C for a time equal to the average life of a B atom, and then the whole would break up. If radioactivity is random, however, they should begin to break up at once and give the usual exponential decay. I do not know whether this experiment has been tried; but I am fairly sure that if the first alternative was correct some related phenomenon would have been noticed.

Radioactivity concerns the nucleus of the atom; most of the evidence about quantum phenomena comes from spectra and other phenomena concerning extra-nuclear electrons. It is harder to find direct evidence for random processes in these. The chief reason for this is that individual occurrences are less violent; consequently, if the phenomena are to be observed at all, so many individual occurrences must take place that the random fluctuation is too small to be measured, and in most cases where random processes might occur the only observable consequence is a continued decay. This is in fact often observed. One instance is the 'canal rays'. A gas at low pressure can be excited by an electric current so that it glows. If it can pass out of the region where the current is passing, the glow continues for a time but gradually fades. This is attributed to the gas being in a state where its atoms carry extra energy; but when left to themselves they dispose of this energy by radiation in characteristic wave-lengths.

The real difficulty is that it is hard to reconcile classical determinism with discontinuous changes. It would apparently require some law that predicts the exact time when every emission of energy takes place. There is no positive evidence for any such law, and, so far as available evidence goes, it supports the view that the ultimate laws concern the probability of a discontinuous change in any interval of time. It is just at this point that quantum theorists differ acutely among themselves, as may be seen from the discussion by Schrödinger[†] and Born.[‡] Schrödinger maintained that the ultimate reality is expressed by the ψ function, which is determinate for all time given its value at one instant; and therefore that classical determinism holds, but refers to ψ functions and not to the positions and momenta of the particles; ψ has no discontinuous changes. Born points out that in a system of N particles ψ is a function of $3N$ co-ordinates and cannot be represented by a distribution over ordinary space; and it is position in ordinary space that we want to know. So far I agree with Born. But Born insists on a 'statistical' interpretation, and this leads to confusion. He interprets the probabilities of wave mechanics as the probability of *finding* the system in a particular state. This leaves it doubtful both what the data are and what the proposition is whose probability is supposed to be stated. In a given experiment on a beam of electrons, say, the probability of catching a particular electron at all is practically zero. If the experiment concerns a gas and is continued until most atoms

† *Brit. J. Phil. Sci.* **3**, 1953, 109–23, 233–42.
‡ *Brit. J. Phil. Sci.* **4**, 1953, 95–106.

have re-emitted radiation, and may therefore be said to have been *found*, it will have been raised to its equilibrium temperature under the radiation; does the probability refer to the initial temperature or to the final one? If the statement deals with what is found, it should apparently refer to the final state. But many of the calculations of quantum theory do not concern the temperature at all, and if they are to be applicable to any particular temperature it appears to be the absolute zero. On the other hand there seems to be no objection in principle to regarding the probabilities as probabilities that the atom, or the system, *is* in specified states. The introduction of *finding* seems to be another unnecessary concession to phenomenalism.

The solutions for an atom or a molecule have to be combined in various ways to give solutions for an assembly such as a gas. In classical mechanics, the fundamental rule for the steady state is the uniform probability distribution of the co-ordinates and momenta; but to a good approximation this can be replaced by independent probability distributions for the various degrees of freedom, provided these are chosen so that the interaction is small. In these the mean energy appears as a parameter.

We may expect an analogue in quantum mechanics. The analogue of the argument for the uniform probability distribution of the co-ordinates and momenta has been developed by Moyal and Bartlett. The independence of the probability distributions no longer holds for indistinguishable particles, so some difference must arise.

Bohr's original formulation of quantum atomic theory spoke of electrons in definite orbits, satisfying the inverse square law of force. The practical difficulty of his theory was that it was impossible to observe the position of an electron in an orbit, so that the details were unverifiable. The further step of the quantum theory, in declaring the orbits to be meaningless, however, strikes me as unjustified. All the hints that I can find seem to suggest that the probability distributions of quantum mechanics are neither based on actual observations nor descriptive of steady states, but are descriptions of systems after disturbance in very special ways. For instance, the result for the ground state of a hydrogen atom, which cannot radiate, gives a positive probability that the separation exceeds any given value. It is difficult to believe that this applies to the atom when quiescent, but it might well be right after disturbance.

Schrödinger's equation takes no account of the finite velocity of light, though it arose from an attempt of de Broglie to do so. A much more general form was developed by Dirac. The probability interpretation clearly needed such an extension, but it is clear that probabilities of events cannot be independent of the observer, when allowance is made for the finite velocity of light, because the region of space-time accessible to A at any instant is not identical with that accessible to B. The probability distribution describing A's knowledge of the world as a function of A's time cannot be transformed into one giving B's knowledge as a function of B's time, for some events known to A are not known to B and vice versa. In fact there is a new uncertainty corresponding to the time of passage of light. Now just at this point present quantum theory is in difficulties; when the interaction of matter and the electromagnetic field is taken into account by successive approximation, divergent integrals are found at the second approximation. I am not prepared to say that the mathematical difficulty arises from the epistemological one, but it is interesting that they appear at the same point.

10·5. Eddington's *Fundamental Theory* maintains that all the laws of physics are not properties of the world, but arise from conventions introduced in the analysis of the data. He has used the illustration of a fishing net; we might infer that all fish are above a certain size, whereas the smallest that we can catch are determined by the mesh of the net that we have chosen. In this case the mesh is determined by the quantum h. Most people would argue that h is not directly measured, but is inferred from a host of observations, but Eddington would reply that the observations depend on solid scales, and that the size of these depends on atomic spacing and hence on quantum considerations. He is right; but (1) the possibility of comparing lengths at all depends on a vast number of observations, besides epistemological principles of inference, (2) Eddington's argument belongs to the realm of physical explanation and not to that of epistemology. In his account of relativity he starts with the notion of a generalized distance between two neighbouring events, but before the notion of such a distance can be considered the whole of special relativity and Newtonian gravitational theory had to be developed to co-ordinate observations, and general

relativity is simply an attempt to combine them into one system.†
Thus his so-called epistemology incorporates most of classical
physics before it even starts. The fishing net analogy is in fact
similar to the point in the account of statistical mechanics given
above (p. 219), since the essence of the matter is that observations
refer to a fixed range of position, they do not follow a particle in
its movement, and measurements are averages. But the size of the
mesh in an actual experiment is not ordinarily determined by h but
is much larger.

Nevertheless it is possible that Eddington's system, though it is
not epistemology, contains much good physics. Unfortunately his
standpoint makes it practically impossible to see what his physical
hypotheses are, and he proceeds by comparing his results at every
point with current quantum theory, which is equally hard to understand. He succeeds in calculating many of the fundamental constants
of physics in terms of three of them, and the agreement with
observation is good. (Rather too good; but it is possible that the
stated uncertainties of the observed values have been inflated.) So
far, however, nobody appears to understand the details of the
calculation. Reference may be made to the work of E. W. Bastin
and C. W. Kilmister,‡ who are studying the algebra used, and
possibly when the mathematics is made clear it will be possible to
see what the physics is.§

Again, if the laws of physics are the result of conventions, we
might ask for the conventions to be stated explicitly. We are entitled
to choose conventions as we like, so long as they serve the purpose
of enabling us to say what we need to say. Here again we do not find
the required statements. To judge by the beginning of *Fundamental
Theory* it might appear that the whole of relativistic wave mechanics
is already adopted as a convention. I think that the actual procedure
is to take over the ideas of more elementary physics—mass, charge,
momentum, energy and so on—and to develop conditions that the
methods of wave mechanics shall be consistent. But whatever the

† The point is developed more fully in *Phil. Mag.* (7) **32**, 1941, 177–205.

‡ *Proc. Roy. Soc.* A, **212**, 1952, 559–75; *Proc. Camb. phil. Soc.* **50**, 1954,
278–86, 439–48.

§ Incidentally I agree with Eddington's remark that the language of current
quantum theory is in such confusion that it is practically impossible to make
a clear statement in it.

extension of old ideas to new subject matter is, it is not a convention. It is a hypothesis, which may be true or false.

One point where hypothesis enters is in the application of the notion of mass to protons and electrons. In elementary measurement mass arises by way of gravity, and to some extent through behaviour of solids in collision. The notion has been extended to molecules; but so far it concerns only bodies that are electrically neutral. To test it for charged particles we should have to compare the attraction between a proton and an electron with the repulsion between two protons; if the laws of gravitation and electrostatics remain true, the magnitudes of the two forces, at equal separation, would differ by about 1 part in 10^{39}. This is far beyond the possibility of measurement.

It is in fact far from obvious that the additive property of mass persists for fundamental particles. In hydrodynamics a solid moving in a fluid produces a motion of the fluid, and when it is accelerated the increase of energy of the fluid has to be taken into account. This requires an increase of the force needed to produce a given acceleration of the solid, and therefore, effectively, an increase in the mass of the solid. Acceleration of a charged particle, similarly, produces changes of the electromagnetic field; this gives an effective increase of inertia, which was used long ago by Larmor to relate the mass of the electron with its radius. We need not examine the correctness of Larmor's model here; the point is the existence of electromagnetic mass. Now the electromagnetic mass of a particle must depend on its surroundings. For instance, near a hydrogen atom the fields of the electron and proton must largely cancel; but in the measurement of the mass of an electron by its acceleration in a known field this effect is much less. What Eddington appears to have done is to find what *constant* masses of the proton and electron will give the correct answers in both circumstances.

At one point he is definitely wrong. Taking the radius of the Universe as R and the number of particles in it as N, he argues that the uncertainty of the position of a general particle is of order R and hence that of the centroid of all particles is of order $R/\sqrt{N}$. This is comparable with the estimated radii of atomic nuclei, and he infers that it establishes a relation between phenomena on the largest and the smallest scales. But the formula $R/\sqrt{N}$ would be right only if the probability distributions for all particles were independent.

Since particles are associated in galaxies, N in the formula should be the number of galaxies, not of particles, and the two scales differ by an enormous factor. Bastin and Kilmister, however, derive the result in another way.

Kilmister and B. O. J. Tupper† give a devastating criticism of Eddington's Statistical Theory,‡ step by step. In their final chapter they rescue some bits that can apparently be put right and get rough agreement with experiment. They have hopes that future refinements may make it better.

In most accounts of quantum theory, at least up to a point, relativity effects are ignored. This amounts to taking the velocity of light as infinite. Now Eddington derives the relation

$$hc/2\pi e^2 = 137,$$

where e is the charge on an electron. If this is a necessity of quantum theory infinite c would imply zero h, and we should be back to classical non-relativistic mechanics.

Eddington speaks much of probability, but it must be said that his treatment is very unsatisfactory. That is not to say that that of his principal physical opponents is any better.

10·6. *Interacting systems.* Consider two systems such that

$$V = V_1 + V_2 + V_{12},$$

where V_1 and V_2 involve the co-ordinates q_1, q_2 of the first and second systems respectively, while V_{12} involves both. If V_{12} was zero Jacobi's equation takes such a form as

$$\frac{\partial S}{\partial t} + H_1\left(\frac{\partial S}{\partial q_1}, q_1\right) + H_2\left(\frac{\partial S}{\partial q_2}, q_2\right) = 0.$$

Evidently with $\partial S/\partial t = -E$ we could write $E = E_1 + E_2$, the partition being arbitrary, and

$$H_1\left(\frac{\partial S}{\partial q_1}, q_1\right) = E_1, \qquad H_2\left(\frac{\partial S}{\partial q_2}, q_2\right) = E_2.$$

† *Eddington's Statistical Theory* (1962). Oxford Mathematical Monographs.
‡ *Eddington, Fundamental Theory* (1946). Cambridge University Press.

Then S will be of the form

$$S = -Et + S_1(q_1) + S_2(q_2).$$

ψ correspondingly will be a product of the ψs for the separate systems. If ψ_{1m}, ψ_{2n} are characteristic solutions of Schrödinger's equation with appropriate E_1, E_2 any form $\psi_{1m}\psi_{2n}$ will be suitable for the joint system. So would any linear combination of such forms if the joint probability density is taken as in p. 232.

If V_{12} is small but not zero the classical treatment is to express it in terms of the α_r and β_r of 10·3 and the time. The α_r and β_r are then treated as variables, and their rates of change are given by new equations of the Hamiltonian form. These are of course small. In quantum theory the parameters given by Schrödinger's equation are treated as constants but their coefficients as variables. There is now a restriction on the forms admissible for the ψ of the joint system. If two particles are similar† in their properties (not including position) the behaviour of the system would be the same if they were interchanged, hence $\psi\psi^*$ would be unaltered. This implies that for the joint system ψ remains unaltered or is simply reversed in sign. This leads to a change in the simple product rule. ψ must be either symmetrical or antisymmetrical in all pairs of similar particles, and the correct combinations for two solutions for similar particles are $\psi_{1m}\psi_{2n} \pm \psi_{1n}\psi_{2m}$. The ψs here are generalized to include the spin. The upper sign given by Bose and Einstein applies to light quanta and helium nuclei, the lower, given by Fermi and Dirac, to protons and electrons. The latter case is known as the exclusion principle. It leads to startling experimental consequences. (1) Two electrons in the same atom cannot have the same quantum numbers. Actually a fourth quantum number arises, according to which an electron carries an intrinsic angular momentum $\pm \frac{1}{2}\hbar$. This explains why if an atom has three electrons only two of them can be in the state of lowest energy; the other must be in a state of higher energy. More complicated atoms have their electrons in shells, those in different shells having different angular momenta. (2) In a crystal the whole crystal must be regarded as the system. The quantum conditions then specify

† The usual word in the literature is *identical*. This would mean that they are the same particle and there is no question of interchange. *Indistinguishable* has been suggested but the possibility of distinguishing their positions remains.

the atomic spacings. Also the outer shells of electrons in an atom may overlap. In metals there is no room for them in their natural places and they wander about through the whole structure. This explains the high electrical conductivity. (3) Two similar atoms can combine by a sort of polarization; two different ones can combine by sharing outer electrons so as to complete a shell. The spin plays an important part here; fuller information is in W. G. Palmer's book.† (4) The spin also affects magnetic properties.

These are only a few of the striking experimental facts that are explained by the theory.

In consequence of the completeness property, however, the systems may interact, the probabilities can still be expressed as linear combinations of those without interaction, but the coefficients will depend on the time. Thus it is possible for the probability to be transferred to new quantum numbers. This accounts for absorption of radiation by an atom, which enters an 'excited' state of higher energy. It may later emit the absorbed energy and fall back into the unexcited state.

There are several possible interpretations of spin. An electron may have size and not be a point charge. (This was suggested long ago by Larmor, who inferred that the mass of an electron should be $\frac{2}{3}e^2/a$, where a is its radius. Heisenberg has revived it.) If it has an orbit with an exact angular momentum the longitude in the orbit, according to the uncertainty principle, should have a uniform probability distribution. The electron could not then be a rigid sphere but would need to be infinitely extensible. A size would make it easier to see how an electron can have a spin angular momentum and a magnetic moment. Alternatively it may be a particle but be oscillating about its expectation position. This appears in Dirac's relativistic theory, in which the velocity oscillates between $\pm$ the velocity of light, and the spin appears automatically.

A misinterpretation of the exclusion principle may be mentioned here. It does not imply that electrons in two different atoms cannot have the same quantum numbers, however far they are apart. To specify the positions of two such electrons the positions of the atomic nuclei are also needed, and these necessarily imply different ψ. The principle however does indicate that two electrons cannot coincide in position for all time, and Eddington

† *Valency* (1944), Cambridge University Press.

thus derives the inverse square law of repulsion as a consequence of the exclusion principle.

10·7. Hartree has shown that if we define, using atomic units,

$$I = \int\int\int \psi^* H\left(-i\frac{\partial}{\partial x_i}, x_i\right)\psi \, d\tau, \quad J = \int\int\int \psi\psi^* \, d\tau$$

through all space, the condition that I shall be stationary $= \lambda J$ for all small variations of ψ subject to J remaining constant is just Schrödinger's equation in the form

$$\nabla^2\psi = 2(V - \lambda)\psi$$

and
$$\lambda\int\int\int \psi\psi^* \, d\tau = \int\int\int \psi^*(-\tfrac{1}{2}\nabla^2\psi + V\psi) \, d\tau.$$

In a conservative system $H = E =$ the energy $=$ constant, and the time enters the solution through a factor e^{-iEt}.

10·8. Schrödinger's equation contains a first derivative with regard to t and second derivatives with regard to the space co-ordinates. A relativistic form for a free particle of the second order in time and space derivatives was found by several authors and is usually known as the Klein–Gordon equation. Dirac[†] succeeded in expressing the operator in this as a product of two operators of the first order, whose coefficients are 4×4 matrices; these operate on a column matrix with four components (a Dirac spinor). This work led to the prediction of the positron, and the spin of the electron appeared as a natural consequence. There have been many further developments.[‡]

I have commented earlier (Section 10·5) on Eddington's extension of Dirac's theory. Dirac's four 4×4 matrices belong to a set of 16, containing 6 anti-commuting pentads. Such an algebra had been given much earlier by W. K. Clifford.[§] Eddington derived the Coulomb law of force as a by-product of the Exclusion Principle and adumbrated an explanation for the value 137 of the constant

† P. A. M. Dirac, *The Principles of Quantum Mechanics*, Chs. 11 and 12, Clarendon Press, 1958.

‡ L. L. Foldy, in D. R. Bates (ed.), *Quantum Theory III, Radiation and High Energy Physics*, Academic Press, 1962. S. S. Schweber, *An Introduction to Relativistic Quantum Theory*, Harper and Row, 1968.

§ *Collected Works* (1882), 181–200, 266–71, 385–94. See also E. A. Power, *The Advancement of Science*, **26**, 1970, 318–28.

hc/e^2. He treated an Einstein universe by the methods of quantum theory and of general relativity; for the consistency of the results he required a definite value of the constant of gravitation, which is consistent with the measured value. He incidentally derived an estimate of the number of particles in such a universe, but he stated that he did not assume that it is the actual universe.

Eddington makes use of a projection $X (X^2 = X)$ of trace 1 and shows that it is of the form

$$X_{ik} = a_i b_k$$

so that X is the product of a column and a row matrix. This may be seen as follows. The eigenvalues of any matrix satisfying $X^2 = X$ are 0 and 1 and there is a non-singular matrix S such that $S^{-1}XS$ is diagonal with elements 0 or 1.† Since the rank and trace of X are equal to those of $S^{-1}XS$ it follows that if the trace of X is 1 there can be only one element 1 in $S^{-1}XS$ and that the rank of X is 1. The requirement that all second order minors must vanish is satisfied by

$$X_{1k} = b_k; \quad X_{ik} = a_i b_k, \quad (i = 2, \dots n, \; k = 1, \dots n).$$

Then the trace is

$$b_1 + a_2 b_2 + \dots + a_n b_n = t, \text{ say,}$$

where we may assume $t \neq 0$.

If b_1 is non-zero and we put

$$a_1 = 1 - (t-1)/b_1,$$

then the matrix X given by $X_{ik} = a_i b_k$ is of rank 1 and of trace 1. The matrix S can be found explicitly. For a (3×3) matrix it is given by

$$S = \begin{pmatrix} a_1 & -b_2 & -b_3 \\ a_2 & b_1 & 0 \\ a_3 & 0 & b_1 \end{pmatrix}.$$

Such matrices play an important part in Eddington's theory. In his usage a_i would be his modification of a Dirac spinor ψ_i and b_k another ψ_k^*.

We may however note a different application. If we consider an experiment that separates a system into several states and X represents a set of probabilities of passage from one state to another $X^2 = X$ would mean that repetition of the experiment does not alter the distribution of probability among the states. If

† D. E. Littlewood. *The Theory of Group Characters* (1940), p. 14. L. Mirsky, *An Introduction to Linear Algebra* (1955), p. 275.

$b_r = a_r^*$, X is hermitian and there is an apparent relation to Yaglom's theorem.

10·9. In earlier chapters I have, I think, made it clear what observed facts constitute strong evidence for the laws derived. In this one I am not clear, nor, I think, is anybody else. Landé quotes seven inconsistent interpretations of the Schrödinger function given by leaders in the subject. I incline to think, with Dirac, that it has no direct physical meaning but is a means of identifying stationary states. What I do insist on is that the phenomena considered require probabilities intrinsic to the system and that understanding needs explicit use of the laws of probability, which are sadly muddled in some accounts. The theorems of Fréchet and Yaglom are probably fundamental.

I should reply to a criticism of the second edition of this book, to the effect that Schrödinger's equation is so complicated that it would have been rejected at the start by the simplicity postulate. The reply is contained in Chapter I, which shows how, in a biological problem, a hypothesis is modified again and again in such a way as to continue to explain things explained by simpler ones, but also to explain new ones. Successive chapters have done this for the more fundamental branches of physics. New parameters are introduced at each stage, essentially by significance tests. I regret to say so, but philosophers, especially philosophers of science, seem unable to attach any meaning to successive approximation.

The argument in this book completely reverses the usual notion of causality. This started with some idea of inherent necessity. After Mach it was replaced by invariable succession, but in fact invariable successions are rare. We now see that science starts with the fact that variation exists and proceeds by first considering the hypothesis that it is random and detecting in succession departures from randomness.

APPENDIX 1

INFINITE NUMBERS

The following remarks are not intended as a full account of the modern theory of infinite numbers, but merely to give some background to some passages in the body of this book. If more is required, G. Cantor's *Transfinite Numbers* or Littlewood's *Elements of the Theory of Real Functions* may be read; a full account is in Whitehead and Russell's *Principia Mathematica*. A glance inside is worth while, as the inside is even more impressive than the outside.

The fundamental notion in number is that of comparison of classes. If we have two classes α and β, such that they can be arranged so that to every member of α corresponds one of β, and vice versa, the classes are said to be equal in number. But if, however we try to pair the members, when all the members of α are paired off there are still some unpaired members of β, then β is said to have the greater number. It was this idea of Cantor that first gave a meaning to infinite numbers; for there is nothing in the notion of comparison that requires the classes to be finite.

The smallest infinite number is the number of the positive integers; this is called $\aleph_0$. We can prove that $\aleph_0$ is also the number of the positive rational numbers. For we can arrange the rational numbers thus:

$$\frac{0}{1}, \frac{1}{1}, \frac{1}{2}, \frac{2}{1}, \frac{1}{3}, \frac{3}{1}, \frac{1}{4}, \frac{2}{3}, \frac{3}{2}, \frac{4}{1}, \frac{1}{5}, \frac{5}{1}, \frac{1}{6}, \frac{2}{5}, \frac{3}{4}, \frac{4}{3}, \frac{5}{2}, \frac{6}{1}, \frac{1}{7}, \frac{3}{5}, \dots$$

Here we first of all group together those rationals with the sum of the numerator and denominator the same, and arrange the groups so that this sum is greater in the fractions of each group than in those of any earlier group. In each group we place the fractions in order of increasing numerator. This arrangement includes every rational fraction, and they are put in a definite order, so that every fraction is reached in a finite number of steps from the beginning. The positive integers 1, 2, 3, ... can therefore be put against them. A one-one correspondence is therefore set up between the rational numbers and the positive integers, and the two classes therefore have the same number.

Classes of number $\aleph_0$ are called enumerable.

Note that the rationals include the positive integers; in this respect infinite numbers differ from finite ones, since an infinite class may have the same number as part of itself.

Other classes with number $\aleph_0$ are: the squares of positive whole numbers; the set of all whole numbers, positive or negative; and the rationals from o to 1. The same is true of the roots of all algebraic equations with rational coefficients. For an equation may be first multiplied by an integer to clear of fractions. For each equation take the sum of the degree and the absolute values of the coefficients; then the equations can be arranged in groups according to the values of this sum, and every equation is reached in a finite number of steps.

The same is true of differential equations of finite order and degree with rational coefficients; we have only to repeat the last argument, first adding the order to the sum of the degree and the absolute values of the coefficients.

This great generality suggests that every infinite class might have number $\aleph_0$. Cantor proved that this is not so. Consider the real numbers from o to 1. Each can be represented by a decimal. Suppose if possible that the decimals from o to 1 can be arranged so that each corresponds to a whole number. For the decimal corresponding to any number n, let the nth figure be a_n. If $a_n = $ o, 1, ..., 7, take $b_n = a_n + 1$. If $a_n = 8$ or 9, take $b_n = $ o. Then the decimal o.$b_1 b_2 ... b_n ...$ differs from every decimal in the enumerated set in at least one place, and hence any such attempt at an enumeration must omit some decimals. (The proviso that if $a_n = 8$ we do not take $b_n = 9$ prevents the new decimal from ending in 9 repeating.)

The set of real numbers from o to 1 is said to have cardinal number c. By analogy with the fact that there are 10^n decimals to the nth place, when n is finite, we can identify c with $10^{\aleph_0}$ or, since we can equally replace 10 by 2 as the basis of enumeration,

$$c = 2^{\aleph_0}.$$

There are infinite numbers larger than c but they do not concern us directly. It is possible that there is no number between $\aleph_0$ and c, but that has not been proved.

Ordinary mathematics does not mention infinity explicitly. It proceeds by attaching meanings to all expressions containing the word *infinity*, but does not define infinity itself. The commonest case is that of a sequence whose nth term is u_n; when we speak of the limit of a sequence as n tends to infinity, we mean that there is a quantity s such that for n large enough all u_n are near s; more strictly, for any positive quantity ϵ, however small, there is an m such that for all n greater than m, $|u_n - s| < \epsilon$. The chief reason for this method is not usually stated, but is as follows. Within the domain of real numbers we can define the fundamental operations of addition, subtraction, multiplication and division. Subtraction of x from y depends on the principle that there is one and only one number in the system which, added to x, gives y. Division of y by x depends on the principle that there is one and only one number in the system that, multiplied by x, gives y; unless $x = 0$, in which case, if y is not 0, there is no multiplier z such that $zx = y$, and if $y = 0$, any multiplier gives $zx = y$. Division by 0 is avoided for the sake of existence and uniqueness; but in the use of 'infinitesimals' it was the curse of the calculus for 200 years.

For infinite quantities addition and multiplication can be defined and used consistently; but they never have unique inverses. If $x = y$ and both are infinite, $y + z = x$ for any z from 0 to x, and $yz = x$ for any z from 1 to x. Consequently the operations of subtraction and division, if applied to infinite quantities, would lead to a host of contradictions.

The number of functions of a real variable is c^c. For to every value of the variable may correspond any of c values of the function, and if the variable can take any real value over an interval there are c values of the variable. Most of these functions, of course, are wildly irregular.

The number of continuous functions is c. For if a continuous function is assigned for all rational values of x it is determined for all real values. Thus the number of continuous functions is not more than

$$c^{\aleph_0} = 2^{\aleph_0^2} = 2^{\aleph_0} = c.$$

It obviously cannot be less than c. The same is true for functions expressible by power series.

If a set of positive numbers x are such that the sum of any finite set of them is $\leqslant a$, where a is fixed and finite, the set is finite or enumerable. Take a sequence of positive numbers $\epsilon_1, \epsilon_2, \epsilon_3, \ldots$ tending to zero ($\epsilon_n = 2^{-n}$ will do). The number of the set of numbers x that satisfy $x > \epsilon_1$ is finite; for there is a number m such that

$m\epsilon_1 > a$, and therefore if the number of this set was infinite we could choose m out of it and have a finite set whose sum exceeds a. Similarly the numbers of x that satisfy $\epsilon_1 \geqslant x > \epsilon_2$, $\epsilon_2 \geqslant x > \epsilon_3$, ..., $\epsilon_{n-1} \geqslant x > \epsilon_n$; ... are all finite. Take them in groups according to increasing n; then every x is reached in a finite number of steps from the start. If the x are finite in number there will be an n such that no more are left; if there is no such n the set is infinite but enumerable.

APPENDIX II

A SUMMATION IN THE
THEORY OF SAMPLING

In 3·4 we needed the sum $S = \sum_{r=0}^{n} {}^rC_l\,{}^{n-r}C_{m-l}$. This is most easily found by a method suggested to me by Dr F. J. W. Whipple. Consider a group of $n+1$ soldiers from whom a set of $r+1$ has to be selected, one to be the leader. (As each is supposed qualified to be the leader, the soldiers may be assumed to be Australian.) Let them be drawn up in one rank, and let the leader be chosen first. Let him be the $(r+1)$th in the rank. Choose l from the men to his left, $m-l$ from those to his right. This can be done in ${}^rC_l\,{}^{n-r}C_{m-l}$ ways; and the sum of these expressions with regard to r is the number of ways of making the selection. But these make up the whole set of ways of selecting $m+1$ men out of the $n+1$; for every selection includes a leader, who must have been in some position in the original rank, and therefore specifies a value of r. Hence

$$\sum_{r=0}^{n} {}^rC_i\,{}^{n-r}C_{m-l} = {}^{n+1}C_{m+1}.$$

The extension to further samples, mentioned in 3·8, can be made as follows. Given that the whole sample of number m is of one type, the probability that the next member taken will be a ϕ is $(m+1)/(m+2)$. If in fact the next member is found to be a ϕ, the probability for the next will be $(m+2)/(m+3)$, and so on. Then by the product rule the probability that m' further members (after the mth) will all be ϕ's is

$$\frac{m+1}{m+2} \cdot \frac{m+2}{m+3} \cdots \frac{m+m'}{m+m'+1} = \frac{m+1}{m+m'+1}.$$

A PROOF THAT π IS IRRATIONAL

Proofs that $\sqrt{2}$ and e are irrational are easy and are given in many books. Hobson's *Plane Trigonometry* contains a rather hard proof that π is irrational. The following was set as an example in the Mathematics Preliminary Examination at Cambridge in 1945 by Dame Mary Cartwright, but she has not traced its origin.

Consider the integrals

$$I_n = \int_{-1}^{1} (1 - x^2)^n \cos \alpha x \, dx.$$

Two integrations by parts give the recurrence relation

$$\alpha^2 I_n = 2n(2n - 1) I_{n-1} - 4n(n - 1) I_{n-2}, \quad n \geqslant 2.$$

If $J_n = \alpha^{2n+1} I_n$ this becomes

$$J_n = 2n(2n - 1) J_{n-1} - 4n(n - 1) \alpha^2 J_{n-2}.$$

Also $\qquad J_0 = 2 \sin \alpha, \quad J_1 = -4\alpha \cos \alpha + 4 \sin \alpha.$

Hence for all n

$$J_n = \alpha^{2n+1} I_n = n!(P_n \sin \alpha + Q_n \cos \alpha).$$

where P_n, Q_n are polynomials in α of degree $\leqslant 2n$, and with integral coefficients depending on n.

Take $\alpha = \tfrac{1}{2}\pi$, and suppose if possible that

$$\tfrac{1}{2}\pi = \frac{b}{a},$$

where a and b are integers. Then

$$\frac{b^{2n+1}}{n!} I_n = P_n a^{2n+1}.$$

The right side is an integer. But $0 < I_n < 2$ since

$$0 < (1 - x^2)^n \cos \tfrac{1}{2}\pi x < 1 \quad \text{for} \quad -1 < x < 1$$

and $b^{2n+1}/n! \to 0$ as $n \to \infty$. Hence for sufficiently large n

$$0 < b^{2n+1} I_n / n! < 1;$$

that is, we could find an integer between 0 and 1. This contradiction shows that $\tfrac{1}{2}\pi$ cannot be of the form b/a.

INDEX